Urban Regeneration an

This book explores the concept of 'home' in Liverpool over phases of 'regeneration' following the Second World War. Using qualitative research in the oral history tradition, it explores what the author conceptualises as 'forward-facing' regeneration in the period up to the 1980s, and neoliberal regeneration interventions that 'prioritise the past' from the 1980s to the present. The author examines how the shift towards city centre-focused redevelopment and 'event-led' initiatives has implications for the way residents make sense of their conceptualisations of 'home', and demonstrates how the shift in regeneration focus, discourse, and practice, away from Liverpool's neighbourhood districts and towards the city centre, has produced changes in the ways that residents identify with neighbourhoods and the city centre, with prominence being given to the latter. Employing Pierre Bourdieu's concepts of habitus and field as mechanisms for understanding different senses of home and shifts from localised views to globalised views, this book will appeal to those with interests in urban sociology, regeneration, geography, sociology, home cultures, and cities.

Clare Kinsella is Senior Lecturer in Criminology at Edge Hill University, UK. Her research interests include cities and urban regeneration, home and homelessness, space and place, policing, fear of crime, and gender, sexuality, and crime.

Routledge Studies in Urban Sociology

This series presents the latest research in urban sociology, welcoming both theoretical and empirical studies that focus on issues including urban conflict, politics and protest, social exclusion and social inclusion, urban regeneration and social class, and the ways in which these affect the social, economic, political, and cultural landscape of urban areas.

Titles in this series

Deindustrialization and Casinos
A Winning Hand?
Alissa Mazar

Urban Regeneration and Neoliberalism
The New Liverpool Home
Clare Kinsella

For more information about this series, please visit: https://www.routledge.com/Routledge-Studies-in-Urban-Sociology/book-series/RSUS

Urban Regeneration and Neoliberalism

The New Liverpool Home

Clare Kinsella

LONDON AND NEW YORK

First published 2021
by Routledge
2 Park Square, Milton Park, Abingdon, Oxon OX14 4RN

and by Routledge
52 Vanderbilt Avenue, New York, NY 10017

Routledge is an imprint of the Taylor & Francis Group, an informa business

British Library Cataloguing-in-Publication Data
A catalogue record for this book is available from the British Library

Library of Congress Cataloging-in-Publication Data
Names: Kinsella, Clare, 1972 – author.
Title: Urban Regeneration and Neoliberalism: The New Liverpool Home
Clare Kinsella.
Description: 1 Edition. | New York: Routledge, 2020. | Series: Routledge Studies in Urban Sociology | Includes bibliographical references and index.
Identifiers: LCCN 2020024038 (print) | LCCN 2020024039 (ebook) | ISBN 9780367861759 (hardback) | ISBN 9781003017363 (ebook)
Subjects: LCSH: Urban renewal – England – Liverpool. | City Planning – England – Liverpool. | Liverpool (England)
Classification: LCC HT178.G72 K55 2020 (print) | LCC HT178.G72 (ebook) |
DDC 307.3/4160942753--dc23
LC record available at https://lccn.loc.gov/2020024038
LC ebook record available at https://lccn.loc.gov/2020024039

ISBN: 978-0-367-86175-9 (hbk)
ISBN: 978-1-003-01736-3 (ebk)

Typeset in Times New Roman
by Deanta Global Publishing Services, Chennai, India

For Mum, Dad, Joey, Pip and, most especially, John

And

In loving memory of John 'Nidge' Duffey

1939–2018

Contents

Acknowledgements

My grateful thanks to all the team at Routledge, especially to Alice Salt who has been supportive and kind beyond measure in helping me produce this text. I would also like to thank Dr Daryl Martin, whose enthusiastic and encouraging review of my proposal played no small part in its acceptance.

I am eternally grateful to my 30 research participants who were good enough to give me an insight into their lives, sharing their highs and lows, their successes and failures, and details of their relationships with their home city. Talking to you all was one of the greatest experiences of my life, and I will be forever thankful.

I should also like to thank Professor Sandra Walklate and Dr Roy Coleman, who supervised the doctoral research that forms the basis of this book, and Dr Lynn Hancock and Dr Phil Boland who examined my thesis. Both Sandra and Roy were exceptionally patient and encouraging as supervisors and provided me with excellent guidance and support throughout the long process. Lynn and Phil, with their insightful comments and questions, gave me confidence in my work. I am incredibly grateful to you all.

I thank my friends and colleagues past and present at Edge Hill University for their kindness, support, suggestions and collegiality: Dr Helen Baker, Dr Alana Barton, Michael Cawley, Julie T Davies, Dr Howard Davis, Dr Alex Dymock, Dr Helen Elfleet, Anita Hobson, Dr Anna Hopkins, Barbara Houghton, Dr John McGarry, Dr Rafe McGregor, Dr Agnieszka Martinowicz, Professor Andrew Millie, Dr Eleanor Peters, Angela Tobin, Alaina Weir, Linda Williams, Dr Holly White, and Joanne White. Extra special thanks to El for allowing me to lean on her at my time of crisis.

My friends and family have been a continual source of support throughout my academic career, and are deserving of particular praise, so a big, heartfelt thank you goes to Sharyn, Bob, Gracie, Phoebe, and Ava Duffey; Zoe, Allan, Amy, and Esme Gibbs-Monaghan; the Gossage family; Alison, Gavin, and Alex Haughton-Muir and all my Chorley/Blackrod in laws; the Kinsella family; Phillipa, Harry, and Jack Malone and Steve 'Coaches' Benson; Joan and Ken McGarry, and Gemma Shiels and Sam Shiels-McNabb.

I must both remember, and thank, my late and much missed parents, Geraldine Kinsella nee Gossage and Tony Kinsella: Dearest Dad, I owe you my nosey nature and both my pride and my curiosity in my home city. Memories of my childhood are filled with visits to family members in the magnificent tenements of Scotland Road and Everton, and long walks around Burlington Street, Vauxhall Road, Soho Street, and William Henry Street – you will always have my love and thanks. Darling Mum, I did all of this for you – your untimely, accidental death 22 years ago caused me to strive to honour you in any way I could. You instilled in me respect for education and told me that I only had to do my best for you to be proud of me; well Mum, I really have done my best for you. I love you as much as ever, and I hope I've done you proud with this book.

Finally, my biggest and most grateful thanks are reserved for my two favourite boys, my husband John McGarry and our son Joe. Johnny Mac, it is safe to say that I would never have been confident enough to try to write a book if it wasn't for your practical help, support, encouragement, and unflinching, endless faith in my abilities – I literally can't thank you enough. Joey Mac, thank you for being the best baby boy in the whole wide world, and for being your cute, cheeky, cuddlesome, comical little self. I love you both more than I can say.

1 Introduction

Regeneration and the making of home in Liverpool

The city of Liverpool is currently enjoying a phase of prosperity and good fortune: A period of renaissance characterised by renewal, prosperity, and success. The 2003 announcement that Liverpool had secured the accolade of European Capital of Culture (ECOC) for 2008 was, for many, confirmation that the city had turned a corner; the bad old days of industrial decline and its various associated problems, and the negative reputations that accompanied them, were well and truly in the past. Contemporarily, Liverpool presents itself as a city on the up – the city centre has been 'made over' as a retail, culture, and leisure hub (Kinsella, 2011) and the city region attracts over 60 million visitors per year, creating a revenue in excess of £4 billion[1] (North West Research, 2019).

The post-ECOC 'bounce' continues, with plans afoot to further develop the waterfront to the north of the city centre, including a new stadium for Everton Football Club at the site of the old Bramley Moore Dock (Liverpool Waters, 2019), and the creation of new, or the resurrection of defunct, railway lines and stations throughout the area (Murphy, 2016). Liverpool is earmarked to benefit from the so-called 'Northern powerhouse' (Parr, 2017), whereby Liverpool and Leeds would mark the outer reaches of a mega city to rival London, incorporating Manchester and having its centre at Hebden Bridge (Davis, 2014). Further, after a 27-year battle, survivors and relatives of the 96 people who died at the Hillsborough disaster in 1989 have finally been rewarded for their efforts with the vindication of the Liverpool fans and the verdict that the victims were unlawfully killed as a result of various police errors (Conn, 2016).

However, just as a coin has two sides, Liverpool has two distinct and opposing images – the face of prosperity, looking outwards towards the rest of the world, and another face, hidden and obscured by the first, which reveals the new, prosperous Liverpool to be a fiction when all factors are considered. Although levels of multiple deprivation in the Liverpool area have gradually been improving since 2004, both Liverpool and nearby Knowsley remain in the top four most deprived local authority areas in England (Ministry of Housing, Communities and Local Government, 2019). Violent crime and, specifically, gun crime, persists as a feature of Liverpool's gang culture and 'county lines'[2] drug markets (Thomas, 2019; McLean et al., 2020), and past events, from political militancy to child criminality, continue to cast a long shadow (Boland, 2008; Furmedge, 2008; Taylor, 2010; Kinsella, 2011; Kelly, 2015; Platt, 2016; Butler, 2020).

New destabilising processes continue to emerge and jeopardise 'new' Liverpool's prosperity, and the status, well-being, and opportunities of people for whom Liverpool is home. The UNESCO[3] world heritage status, ascribed to the "Maritime mercantile city" (Gaillard and Rodwell, 2015, p24) of Liverpool in 2004, has been under threat since 2006 due to the World Heritage Committee's "serious concerns" (ibid.) about new buildings erected at the Pier Head deemed unsympathetic to the existing skyline. Only in June 2019, UNESCO requested a moratorium on further large-scale development, threatening that Liverpool "could lose its highly prized heritage status if it fails to comply" (Kirby, 2019). Meanwhile, housing in Liverpool is "in crisis" (Parry, 2016) as "only one new home is being built for every EIGHT new arrivals in Liverpool" (ibid., emphasis in original), yet there are currently thousands of empty properties in the Liverpool area (Liverpool City Council, 2016), most of which are in populated and settled streets, though in some cases entire streets have been empty for several years (Foster, 2016). The emergence of the 'gig economy' based on the rise of the zero hours contract (Standing, 2011; Getting By?, 2015; Snider, 2018; Koumenta and Williams, 2019), together with restrictions on mortgage provision since the global financial crash of 2007/2008 (Scanlon et al., 2011) mean that securing a home in Liverpool is becoming increasingly more difficult. Already high rentals increase every six months as properties are revalued, tenancies are short, and landlords can be everything from absent and disinterested to deceitful and criminal (Cosslett, 2015; Getting By?, 2015; Paton and Cooper, 2017).

This book is about Liverpool and home. More specifically, the book focuses on the experiences of home within the Liverpool setting, and how they have been moulded by the changing fortunes of the city, particularly in relation to the various regenerative processes that Liverpool and its people have been exposed to. After the Second World War, much of the built environment in Liverpool has been subject to some form of regeneration, whether as a result of 'slum'[4] clearance, road system restructuring, or more nebulous processes such as renaming or repackaging of an area or phenomenon in a bid to change attitudes and/or image (Leeson, 1970; Muchnick, 1970; Broomfield, 1971; Topping and Smith, 1977; Wilson and Womersley, 1977; Pooley and Irish, 1984; Hayes, 1987; Couch, 2003; Balderstone et al., 2014). Contemporarily, regeneration initiatives in Liverpool, conceived very much in the fading glow of ECOC 2008, have undergone a shift away from physical regeneration of the city towards attempts to regenerate the city's image, reputation, and standing in comparison to other cities. Recent trends have been towards a cultural revival of the city, with an emphasis on Liverpool's maritime heritage and the many historic buildings in the area, particularly in the city centre. Put very simply, the main thrust of the argument presented is that changes in regeneration emphasis over time, moving from a focus on the districts – inner areas to overspill estates and new towns – to a focus on the city centre (documented by Coleman, 2004) and, most specifically, the waterfront, have led Liverpool people to conceptualise 'their' Liverpool home in the contemporary context as synonymous with the city centre rather than the districts where they live, or used to live.

Thinking about home

Home at first sight appears as a very straightforward concept; everyone knows what a home *is*, and everybody knows what having a home *means*. Yet home as a concept is deceptive; its surface simplicity hides a very deep and complex abstraction, which is multi-layered, multi-faceted, and interpreted in different ways contingent on a range of factors including time, place, epoch, culture, geographical setting, privacy, and power relations. In short, home is a social construction – an abstract idea rather than a uniform concept which manifests itself in the same way at all times (Tuan, 1977; Dovey, 1985; Rybczynski, 1986; Wright, 1991; Somerville, 1992; Massey, 1994a, 1994b; Wood and Beck, 1994; Waghorn, 2009). Born from the human need for both shelter and roots, home is variously understood to mean a place to 'be', that is, a place where one can 'house' their human existence (Rykwert, 1991; Kearon and Leach, 2000), belonging (Windsong, 2010), acceptance (Kidd and Evans, 2011), security (Despres, 1991; Waghorn, 2009), memories (Dovey, 1985), and rules (Douglas, 1991; Morley, 2000). However, it is important to recognise that home operates on different levels – home can mean the dwelling within which one lives; equally, though, it can mean the immediate environs or neighbourhood surrounding one's dwelling, hometown, or home nation (Hayward, 1977; Morley, 2000; Lewicka, 2005, 2010; Manturuk et al., 2010; Mac an Ghaill and Haywood, 2011). Depending on a range of variables, home can connote feelings of both 'having space' and 'being placed', and be representative of freedom, imprisonment, or both (Morley, 2000; Mallett, 2004; Waghorn, 2009).

This book takes the concept of home and locates it within a Liverpool context. All aspects of home in Liverpool, whether related to dwelling, neighbourhood or hometown, feelings, experiences or memories, have been touched by regeneration initiatives and processes in one way or another. Some people for whom Liverpool is home have experienced significant impact on their versions of home via these processes; others meanwhile experience them in a more abstract and opaque way. From an exploration of the literature on home as a concept, and the literature on Liverpool's sometimes tumultuous relationship with regeneration, I have been able to conceptualise 'regenerating Liverpool' into two distinct, yet overlapping, eras. The first identifiable era, from the end of the Second World War to the late 1970s/early 1980s, I have termed the 'forward-facing era'; a period when regeneration focused on the future, based on the belief that Liverpool would continue to develop and prosper and, consequently, needed to be 'modernised' to cope with the demands of an increasing population, a strong workforce, and a buoyant, flourishing economy. The second identifiable period I have termed the 'backward-facing era'; emerging towards the end of the 1970s, a disastrous decade for the city, this era represents an about face in terms of understandings of where prosperity lies for Liverpool, and a 180-degree change of direction from prosperity based on an insecure, uncertain future, to prosperity based on the city's past – its history, heritage, and status as maritime city of empire.

To summarise: The post-war period up until the late 1970s/early 1980s I have characterised as forward facing, due to its optimism and focus on the future; the period from the late 1970s/early 1980s I have characterised as backward facing, due to its prioritisation of the past. Thus, regeneration initiatives during the first era are understood to be anticipating a future whereby Liverpool remains on the same trajectory as it always has been – pushing forward and modernising to create a city ready for the 21st century. Regeneration initiatives during the backward-facing era, however, abandon the existing trajectory because of continued set-backs and false dawns, and adopt a new trajectory which relies heavily on the certainties of the past as a resource which can be continually mined for prosperity-creating history, legacy, heritage, and culture.

The *volte-face* from focus on the future to prioritisation of the past arguably emerged with the restoration of the derelict, historic Albert Dock, kick-started by the attentions of the 'Minister for Merseyside', Michael Heseltine, in the early 1980s (Crick, 1997; Couch, 2003; Avery, 2007; Taylor, 2010). Contemporarily, though, backward-facing regeneration manifests itself predominantly as event-led regeneration, which relies on the premise that 'mega-events' like, for example, the London Olympics of 2012, will bring with them a regeneration 'boost' to the location where they are held and to the local population (Sadd, 2010; Edizel, 2014). Because of the identifiable focus on the past that characterises aspects of contemporary regeneration, commemoration of past events and past glories is a common basis for events in Liverpool – since ECOC 2008, there have been a range of commemorative events tagged to some aspect of the city's history. Sea Odyssey, for example, brought the 'little girl giant puppet' to the city centre to mark the centenary of the sinking of the Titanic (Fraser, 2015); the 'Three Queens', saw three liners sail up the River Mersey to commemorate 175 years since the birth of the Cunard cruise line (Longman, 2016); and '50 Summers of Love' was a programme of events to mark, well, the 50th anniversary of 1967 (Culture Liverpool, 2017). To the cynical, each commemorative event seems more contrived and phoney than the last, yet such events are evidential of certain phenomena crucial to my central argument – all such events are set predominantly at the city centre, and, more specifically, the waterfront; supporting the notion that contemporary regeneration initiatives prioritise the city centre over the districts.

These combined issues were distilled into two key research questions which underpinned the doctoral research that the book is based upon:

1. What impact has the 'forward-facing' regeneration of the 1950s–1970s had on experiences of and feelings about the Liverpool home?
2. What impact has the 'backward-facing' regeneration of the 1980s onwards had on experiences of and feelings about the Liverpool home?

Primary research was conducted for the project with 30 people to whom Liverpool is home in one way or another – some are Liverpool residents, and have been since birth; others now live elsewhere, but Liverpool retains a certain significance to them. For contextualisation purposes, an account of the research design and research process can be found below.

Methodology

Learning is both more achievable, and more desirable, in social science research than *proving*. The primary research that forms the basis of this book, which consists of the testimony of 30 people for whom Liverpool is some form of home, was wholly qualitative in nature, and, given the emotional and personal nature of the topic area, driven by the desire to prize subjectivity over objectivity. Rejecting any notion of the search for impartial facts in the positivist tradition, the research was conceived very much as an interpretivist endeavour, prioritising individual experience, meaning, understanding, and personal 'truth'. Further, the research was conducted from a particular standpoint; rather than clinging to any façade of neutrality, which can never in reality be achieved, from the outset I have overtly been on the 'side' of my respondents (Harvey, 1990).

In valuing the perspective of my respondents and their subjective knowledge, I have been influenced by the approach to knowledge outlined by Michel Foucault (1972, 1976). In *The Archaeology of Knowledge* (1972), Foucault employs the term 'archaeology' to refer to the uncovering or (re-) discovery of knowledge and 'truth' which has been hidden or unheard due to domination by other, standard, validated knowledge. Later, in *Power/Knowledge* (1976), Foucault elaborates further on the subject of knowledge and, in particular, what he terms "subjugated knowledges" (p346). Developing the notion of uncovering knowledge via archaeology, Foucault uses the term "genealogy" (p347) to describe the process of looking back through history to trace the development of knowledge, particularly knowledge of historical struggles which have been buried. This buried 'truth', together with knowledges that have been disregarded, constitute subjugated knowledges; under-recognised, undervalued, and unheard. Subjugated knowledge is more meaningful in revealing the true nature of the social world, and the operation of power, precisely because it is distinct from obvious, commonsense, scientifically proven knowledge. The Foucauldian approach of this project facilitates the unearthing of the subjugated knowledge that questions and challenges the narrative of a regenerated Liverpool, as established by local government, national government, and other primary definers with a vested interest in Liverpool's image and prosperity.

In terms of the research methods deployed, the research design drew heavily on the narrative/oral history tradition in terms of data collection, and the grounded theory method for data analysis. In very simple terms, narrative research entails excavating knowledge, often in the form of a counter-narrative, which takes a different perspective to, or even challenges, accepted knowledge (Howarth, 1998; Aston, 2001; Cole and Knowles, 2001; Rose, 2001; Fielding, 2006; Gardner, 2006; Reed-Danahay, 2006). In particular, life history and oral history have a clear focus on shifting the power dynamics of research by giving voice to those whose views are seldom privileged – indeed the process of recounting pertinent life events from an individual perspective is prized as being an enriching, empowering, and cathartic experience in and of itself (Howarth, 1998; Aston, 2001). Thus, life history research provides more than a means of simply preserving something of ourselves for posterity[5] (Janesick, 2007).

There are clear similarities and overlaps between narrative methods and oral and life history; however, there is one device that is central to all three – the concept of *story* (Emihovich, 1995; Polkinghorne, 1995; Howarth, 1998; Cole and Knowles, 2001). Polkinghorne (1995) unpacks the everyday notion of story by pointing to the self-editing that protagonists engage in when formulating their stories, highlighting key plot points and "inciting incidents" (Woodside, 2010, p49) – a process which facilitates both sense-making of events, or "storied knowing" (Polkinghorne, 1995, p9) at the individual level, and the dramaturgy[6] of external presentation (Flyvbjerg, 2006; Woodside, 2010). Here there is a recognition that narrative and life history are essentially social and cultural constructions (Polkinghorne, 1995; Fielding, 2006), not least because memory also shares these characteristics (Woodside, 2010). However, that is not to say that the notion of story within narrative and life history research is flawed simply because it relies on individual memory and is difficult or impossible to verify[7] – Emihovich (1995), for example, states that "stories do not pretend to be objective because they deal with emotions, the irrational part of behaviour, they tap into qualities of imagination and fantasy" (p39). Further, Zeller (1995) discusses the interface between fact and fiction which, by necessity, has crept into narrative research as a means of stressing the gravity of social events as experienced by humans without stripping the narrative of its humanity by reporting it in a bland, factual, dispassionate way.[8] Thus, the inherently subjective nature of story, and the subsequent mismatch between story and objectivity, is by no means a barrier to presenting stories in social science research, not least because stories invoke passion; without passion, researchers are "desiccated intellectuals" (Emihovich, 1995, p43).

Whilst the oral/narrative history method enabled me to gather my data, it is the principles of the grounded theory method that have shaped my analysis. Developed in the 1960s by Glaser and Strauss as a challenge to the view that all possible theories have already been established by the "great men of sociology" (Urquhart, 2013, p15), grounded theory method is an attempt to empower individual social researchers by facilitating the development of new theories grounded in evidential data (Glaser and Strauss, 1967). Starting with the premise that social research places too much emphasis on verification and validation, Glaser and Strauss sought to 'free' researchers from the constraints of existing 'knowledge' and the inherently hierarchical nature of academia, thereby creating a democratising method which seeks to undermine an academic system where elites engaging in "grandiose armchair theorising" (Dey, 2007, p172) remain forever at the top and academic mobility is limited.

Grounded theory in its purest form requires researchers to start with a 'clean page', having not read any of the literature already written on the subject area (Holton, 2007; Lempert, 2007; Birks and Mills, 2011; Urquhart, 2013). However, many grounded theory method proponents recognise that such an approach is unrealistic in modern academia, leading Urquhart (2013, p11) to suggest that "researchers should have an open mind, as opposed to an empty head". Once data has been collected, the analysis process begins with the "constant comparative method" (Harding, 2006, p131), whereby the first 'piece' of data is compared to the second, the second compared to the third, and so on, to identify commonalities. This systematic 'coding' of 'categories' as data is received allows the researcher to become immersed in the data and, crucially, encourages an analysis of the data to emerge which incorporates an element of the researcher, under the researcher's control: "The analyst can continually adjust his [sic] control of data collection to ensure the data's relevance to the impersonal criteria of his [sic] emerging theory" (Glaser and Strauss, 1967, p48). The data collection should only cease when the data has become 'saturated', that is, when no new data is being generated and existing data is merely being reinforced – this is the time to stop (ibid.).

Participant recruitment occurred organically, and in concentric circles radiating out from me both as a starting point and as gatekeeper. I treated my first interview, with my closest friend, Mary, very much as a pilot; testing the water with questions and some 'kite-flying' in terms of topic areas. One of the issues that came up during this interview was the Kensington New Deal for Communities, and Mary suggested some people who would be interested in talking to me about it. She sounded out Rita and Irene on my behalf, and they both agreed to be interviewed. This is an example of the data dictating where I should sample next, in keeping with the grounded theory method (Glaser and Strauss, 1967; Urquhart, 2013), which happened several times throughout the data collection process.

Other participants were selected for different reasons. Some, like Jill and Martha, knew of my project and volunteered to take part. John heard of my project via his daughter-in-law, a close friend of mine, and persecuted her until I went to interview him and his wife Anne. I approached Jodie because of her experience of 'third-sector' work in the Lodge Lane area, and Susie because of the various outreach roles she has had over the years. I heard Ronnie Hughes speak at a one-day conference about the plight of Eldon Grove; a dilapidated, but listed, tenement building in the Scotland Road area. He mentioned his time working as a housing officer in inner Liverpool; I was able to get his contact details and sent him a speculative email, and he agreed to be interviewed. Several of my extended family members live abroad, and my cousins Dave, Joey, and Sandra all agreed to complete a questionnaire via email, thereby adding an additional dimension to my data. Colette's sister in law, Enid, lives in America; she agreed to complete the questionnaire but, on reading it, decided she would prefer to provide a statement, which I have used as a testimony (Polkinghorne, 1995).

In terms of presenting the data collected, I took the decision to use first names only[9] as (a) a full name provides a greater opportunity for identification of individuals and (b) I felt that being on 'first-name terms' would enhance the qualitative nature of the study and facilitate a closeness between the research participants and any potential readers – I wanted my respondents to come across as 'real' people, and their voices to be heard and valued. I offered every participant the option of either using their own name, reflecting the notion of empowerment through the 'owning' of testimony' (Kelly et al., 1994); some opted for a pseudonym, some preferred their own name – although there are two cases where I took the decision to change names as I felt that the use of a real name, together with detail provided in the testimony, might lead to identity being exposed (ibid.). Three names were changed simply to avoid confusion; some respondents shared the same name, others had a name used in the narrative.

Data gathered via either interview, questionnaire, or written testimony reveals the project's chief finding: When my respondents were asked to think about the relationship between home and Liverpool in relation to the *forward-facing* period of regeneration, roughly, 1945–1980, they almost universally interpreted and understood home to be their *neighbourhood* – the section of Liverpool, whether this be a named district, an estate, or a small collection of streets – which immediately surrounds/surrounded their dwelling. However, when respondents were asked to think about the relationship between home and Liverpool in relation to the *backward-facing* period, from around 1980 onwards, they almost universally interpreted and understood home to be the *city centre*; more particularly, the *waterfront*.

Liverpool, *habitus* and home

The concept that I have explored, as a possible explanation of this shift of understandings of regeneration and home in Liverpool from the micro/meso to the macro, is the Bourdieusian concept of *habitus* – the notion that groups of people can be brought to a particular view, slowly and surreptitiously, which is entirely different to their previous perspective, without them even perceiving that their view has changed (Bourdieu, 1977, 1983, 1985). Stealthy and insidious, *habitus* is an unseen and unacknowledged process whereby a population are, subconsciously or with minimal consciousness, brought to a position, via subtle and subliminal signals and prompts over a long period of time. The success of *habitus* lies within its implicit nature; change is encouraged so slowly and gently that those subject to said processes of change do not notice the difference, or even that they are being manipulated, just like the proverbial frog in water, heated up so gradually that it does not realise it is boiling to death (Throgmorton, 2003). Habitus as a concept is employed here as a mechanism for *making sense of* what has been found.

Liverpool is my home city, and I have lived within it all my life, other than the five years I spent in nearby Sefton. As such, the project is very personal to me, and it represents both an academic endeavour and a political, critical, 'view from below' presentation of the experiences of regeneration processes of Liverpool people. In my forties, I fit the demographic of my research sample perfectly, and many of the events and phenomena that have impacted on the lives on my respondents have also touched my life to a greater or lesser extent. Thus, I am fully entrenched and immersed in this project, in every sense. Whilst this level of subjectivity is perfectly valid and appropriate (Harvey, 1990; Reed-Danahay,

2006), it has also given me cause for concern; Roth (2009) notes that auto-ethnography can result in self-indulgence and even narcissism when done badly. Further, as a Liverpudlian I am subject to the same prejudices, reputations, and stereotypes as all others, and I am not immune to criticisms levelled at the city and its population. In my wide and multi-disciplinary reading on Liverpool and various facets of Liverpool life, I have developed an awareness that different standpoints are taken when writing about/analysing/theorising Liverpool and its people. Specifically, I have identified four loose genres of standpoint on Liverpool that have helped me to understand the body of literature available, and also my position in relation to it.

Aside from the easily identifiable, commonplace, stereotype-heavy commentary on Liverpool, which largely reflects the views and perspectives of people looking at Liverpool from the outside, and which I term simply as the 'negative' standpoint,[10] three others can be identified which are more reflective of commentary and analysis from within the city. The first, which I have termed the 'plucky' Scouser[11] standpoint, is somewhat sentimental; dwelling on such supposedly universal Scouse qualities as resilience, cheery buoyancy in the face of adversity, and the ubiquitous good sense of humour; the tone of literature from this standpoint is characterised by pride in the city and its people and optimism for the future. I have encountered several texts in this vein, and have become frustrated as, whilst they are very useful for factual and historical accounts of events, they are spoiled for me by their spirit of bravery, persistence, and 'we always bounce back' mentality, which diminishes the negative effects of adverse events and processes.

Conversely, I am much more comfortable reading literature on Liverpool which acknowledges the various, perhaps unique, problems that the city has faced, and the harms that have resulted from them. This 'mardy'[12] perspective, as I have termed it, is inherently critical in that, rather than accept the 'top-down' views from outside of the city, which 'plucky' Scouse writers have colluded with by focusing on change within the city and Liverpool people taking responsibility for their own plight, it focuses on the structural wounds that have been inflicted on the city. The 'mardy Scouse' perspective, then, challenges the perception that Liverpool people are only harmed by negative social processes in a minor way, if at all, because of their inherent resilience and jocular disposition. By shining a light on the social, political, economic, and cultural processes which have conspired to simultaneously wreak harm on Liverpool, and attribute blame for said harm on the Liverpool populace, this 'mardy' standpoint recognises Liverpool's structural location as a city and seeks to ameliorate it.

The final position that I have identified in relation to Liverpool, which I have termed the 'everything is awesome' standpoint, has emerged more recently and is reflective of the cultural turn that renewal and regeneration have taken in the Liverpool context. Titled as an homage to *The Lego Movie*, this standpoint somewhat disregards the ongoing problems of the city in favour of prioritising prosperity, renaissance, and the social cohesion and civic pride that are thought to characterise contemporary Liverpool. Within this genre of standpoint, commentators are effectively cheerleaders for the city who seek to both contribute to, and perhaps benefit from, Liverpool's post-ECOC climate of cultural engagement, creative industries, and the varying marketing strategies which support them. The 'everything is awesome' perspective presents a one-dimensional picture of contemporary Liverpool; a picture which denies the realities of some Liverpool lives.

It is important to be cognisant, when reading literature, that it is never created in a vacuum; it is always reflective of the motivation for writing it, the position of its author(s), the perceived position of its audience, and the prevailing social/cultural/political/economic climate of the time. As I have had to be mindful of the context within which existing literature has been written, it has also been essential that I keep in mind both my structural location and my standpoint when conducting primary research. Thus, whilst I view my position as being reflective of the 'mardy' standpoint, if any, I have to accept that what I have produced may be interpreted in a different way; my entrenchment in this project and the city upon which it has been conducted could possibly be viewed in some quarters as sentimental subjectivity.

Overview of the book

The book comprises seven substantive chapters framed by this introduction and a conclusion. Chapter Two provides a multi-disciplinary, theoretical examination of the concept of 'home'; what it is, where it has come from, and what it means from various perspectives. To provide a greater understanding of the concept, the state of being without a home – homelessness – is also considered, and the phenomenon which has come to be known as domicide – loss, or destruction, of home – is conceptualised towards the end of the chapter.

Chapters Three, Four, and Five essentially comprise a review of literature on the city of Liverpool, its people, and the various phenomena, processes, and events that have impacted on both throughout the 20th and 21st centuries. Whilst Chapter Three provides 'scene setting' with detail on Liverpool life prior to the Second World War (WWII), and long-standing, entrenched divisions and harmonies between the Liverpool populace, Chapter Four explores the time period between, roughly, 1945 and 1980, conceptualised as the 'forward-facing era' in my analysis. The 1950s/1960s 'boom' – characterised by a buoyant economy, an emerging popular culture, 'slum' clearance, 'modernisation', and optimism – is explored, followed by the 'bust' of the 1970s, which brought with it recession, economic decline, unemployment, urban decay, and the death of hopes for brighter days in a 'futuristic' Liverpool. The chapter closes with a consideration of the emerging negative discourses on the city and its inhabitants, and how they were challenged by a new, authentic, working-class Liverpool 'voice' via the media of literature, theatre, and television.

Chapter Five focuses on what I have termed the 'backward-facing' period, broadly from the late 1970s/early 1980s to date. Taking as its starting point the realisation that Liverpool's future prosperity lies in the past, this chapter details nascent ideas of culture and heritage as resources to be harnessed for profit, contextualised against the turbulence and perceived criminality of the 1980s, and the efforts to turn around 'problem Liverpool' via Objective 1 funding and New Labour policy in the 1990s. Regeneration and home in millennial Liverpool are located against processes of consumer- and culture-led regeneration which rely both on the governance and statecraft of 'merged' local government and 'mega-investors' such as Peel Holdings and Grosvenor Development. Significantly, the chapter highlights the contrast between focus on housing regeneration in the districts during the 'forward-facing' period and the cultural-/heritage-driven regeneration of the city centre and waterfront that characterises the contemporary 'backward-facing' period. This chapter closes the literature review section of the book.

Chapters Six and Seven present my data findings and analysis. In keeping with the notion of 'forward-facing' and 'backward-facing' periods of Liverpool regeneration, Chapter Six presents data and analysis pertinent to the former; Chapter Seven the latter. The codes generated from the grounded theory method data analysis are 'scaled up' into themes which provide a structure for each chapter and, in keeping with the principles of both the critical standpoint and the oral/life history method, direct quotes from respondents are used frequently to prioritise their voice over mine. From the analysis in these two chapters emerges, in Chapter Eight, the project's chief finding – regeneration and home during the 'forward-facing' era leads to respondents focusing on dwelling and neighbourhood; regeneration and home during the 'backward-facing' era appears to create more focus on the city centre and, most specifically, the waterfront.

The book closes with a Conclusion which provides direct responses to the key research questions, reflects on both the implications for findings and the research process itself, and potential areas for further research and analysis. Pen portraits of my respondents are included as appendices.

Notes

1 In 2019, 67.38 million visitors came to the Liverpool city region and spent an estimated £4.93 billion (North West Research, 2019).
2 The term 'county lines' has come to refer to the supply of drugs in small towns and rural areas from bigger, well-established drug markets such as Liverpool, and often involves the exploitation of young and vulnerable people (Robinson et al., 2019).
3 "The United Nations Educational, Scientific and Cultural Organization (UNESCO) seeks to encourage the identification, protection and preservation of cultural and natural heritage around the world considered to be of outstanding value to humanity. This is embodied in an international treaty called the Convention concerning the Protection of the World Cultural and Natural Heritage, adopted by UNESCO in 1972" (UNESCO, n.d.).
4 Gilbert (2007) demonstrates the myriad problems associated with the term 'slum', not least that it is disputed, culture-specific, and too readily used to attribute negative qualities to associated populations. Throughout the book I present the word in inverted commas to indicate its contested and problematic nature.
5 See e.g., Ehlman et al. (2014) for a discussion on the empowering properties of engaging with oral history projects amongst elderly populations.
6 Reflecting Goffman's (1996) work using the analogy of the theatre to understand social activity as a performance.

7 Polkinghorne (1995, p7) notes that "*story* carries a connotation of falsehood or misrepresentation".

8 Zeller (1995) points to the emergence of 'new' or 'literary' journalism during the Vietnam War which adopted the method and style of fiction writing to impress on the reader the human horror of events, as opposed to presenting facts in an objective, dispassionate manner.

9 The only exception is Ronnie Hughes who, as a community activist, wanted his name associated with comments made to facilitate and augment his activism and current projects.

10 Much has been written, particularly within the news media, about 'problem' Liverpool and 'problem' Liverpudlians, over the years. The tone of such material is often one of contempt and disdain, an example being Beryl Bainbridge's 1984 work, *English Journey, or, the Road to Milton Keynes*; Ms Bainbridge, herself born in nearby Formby to a Liverpool family, is particularly scathing of the perceived actions of Liverpool people and a once great city ravaged to the extent that "it's as well the Liver Birds are tied down with cables otherwise they'd fly away in disgust" (p94).

11 Derived from the name of stew-based meal and used to define a local accent, a dialect, and even a world view, 'Scouse' is the well-known nickname for people who hail from the Liverpool area (Crowley, 2012). The notion of a Scouse identity is a complex and contested one (see e.g., Boland 2010), and the moniker is accepted by some Liverpool people and rejected by others. I use the term here as a signifier of the acceptance of the 'plucky' narrative which, perhaps, lacks the critical and politicised edge of other positions.

12 The colloquial term 'mardy' appears to have a disputed origin; the Oxford English Dictionary holds that the term is a bastardisation of 'marred', meaning spoilt, as in a spoilt child, whilst the online Urban Dictionary states that it is a variation on the adjective 'moody'. It is interpreted variously as sulky, irritable, grumpy, bad-tempered, terse, petulant and, perhaps most pertinently given some criticism in recent years of Liverpudlians as being mawkish and badly done to, as a description of a cry baby. Within my own orbit, 'mardy' has always been an indicator of objection to an imposition or an unfortunate circumstance, accompanied by frustration; babies 'giving out' i.e., shouting or crying because they are teething, or perhaps uncomfortable in hot weather, are pronounced to be 'mardy', as are people who object vocally to something when others would prefer that they be quiet. I have selected 'mardy' as an appropriate term here as it encapsulates a sense of vocalising the negative and the undeserved.

Bibliography

Aston J (2001) 'Research as relationship' in Cole A and Knowles J (eds) *Lives in Context: The Art of Life History Research*. Alta Mira: Oxford, pp 145–151.

Avery P (2007) 'Born again: From dock cities to cities of culture' in Smith M (ed) *Tourism, Culture and Regeneration*. CAB International: Wallingford, pp 151–162. doi:10.1079/9781845931308.0000.

Bainbridge B (1984) *English Journey, or, the Road to Milton Keynes*. Carroll and Graff: New York, NY.

Balderstone L, Milne G and Mulhearn R (2014) 'Memory and place on the Liverpool waterfront in the mid-twentieth century' *Urban History* 41 (3) pp 478–496. doi:10.1017/S0963926813000734.

Birks M and Mills J (2011) *Grounded Theory: A Practical. Guide*. Sage: London.

Boland P (2008) 'The construction of images of people and place: Labelling Liverpool and stereotyping Scousers' *Cities* 25 pp 355–369. doi:10.1016/j.cities2008.09.003.

Boland P (2010) 'Sonic geography, place and race in the formation of local identity: Liverpool and Scousers' *Geografiska Annaler Series B Human Geography* 92 (1) pp 1–22. doi:10.1111/j.1468-0467.2010.00330.x.

Bourdieu P (1977) *Outline of a Theory of Practice*. Cambridge University Press: Cambridge.

Bourdieu P (1983) 'The field of cultural production, or: The economic world reversed' in Johnson R (ed) (1993) *The Field of Cultural Production: Essays on Art and Literature*. Polity: Cambridge, pp 29–73.

Bourdieu P (1985) 'The genesis of the concepts of habitus and field' *Sociocriticism* 2 pp 11–24.

Broomfield N (1971) *Who Cares?* Documentary. Available at http://www.liverpoolpictorial.co.uk/blog/liverpool-1971/. Accessed 30 March 2016.

Butler A (2020) 'Toxic Toxteth: Understanding press stigmatisation of Toxteth during the 1981 uprising' *Journalism* 21 (4) pp 541–556. doi:10.1177%2F1464884918822666.

Cole A and Knowles J (2001) 'What is life history research?' in Cole A and Knowles J (eds) *Lives in Context: The Art of Life History Research*. Alta Mira: Oxford, pp 9–24.

Coleman R (2004) *Reclaiming the Streets: Surveillance, Social Control and the City*. Willan: Cullompton.

Conn D (2016) 'Hillsborough inquests jury rules 96 victims were unlawfully killed' *The Guardian*, 26 April 2016. Available at https://www.theguardian.com/uk-news/2016/apr/26/hillsborough-inquests-jury-says-96-victims-were-unlawfully-killed. Accessed 22 June 2016.

Cosslett R (2015) 'Generation Rent v the landlords: "They can't evict millions of us"' *The Guardian*, 22 August 2015. Available at https://www.theguardian.com/global/2015/aug/22/renters-fight-rogue-landlords-rhiannon-lucy-cosslett. Accessed 8 July 2017.

Couch C (2003) *City of Change and Challenge: Urban Planning and Regeneration in Liverpool*. Ashgate: Aldershot.

Crick M (1997) *Michael Heseltine: A Biography*. Penguin: London.

Crowley T (2012) *Scouse: A Social and Cultural History*. Liverpool University Press: Liverpool.

Culture Liverpool (2017) *67–17: 50 Summers of Love*. Available at https://www.cultureliverpool.co.uk/summer-of-love/. Accessed 8 May 2017.

Davis E (2014) 'The case for making Hebden Bridge the UK's second city' *BBC News*, 10 March 2014. Available at http://www.bbc.co.uk/news/business-26472423. Accessed 5 July 2017.

Despres C (1991) 'The meaning of home: Literature review and directions for future research and theoretical development' *Journal of Architectural and Planning Research* 8 pp 96–115.

Dey I (2007) 'Grounding categories' in Charmaz K and Bryant A (eds) *The Sage Handbook of Grounded Theory*. Sage: London, pp 167–190. doi:10.4135/9781848607941.n8.

Douglas M (1991) 'The idea of a home: a kind of space' *Social Research* 58 (1) pp 287–307.

Dovey K (1985) 'Home and homelessness' in Altman I and Werner C (eds) *Home Environments*. Plenum: New York, NY, pp 33–64.

Edizel H (2014) *Governance of Sustainable Event Led Regeneration: The Case of London 2012 Olympics*. Doctoral thesis, Brunel University. Available at https://core.ac.uk/download/pdf/20666120.pdf. Accessed 9 July 2017.

Ehlman K, Ligon M and Moriello G (2014) 'The impact of intergenerational oral history on perceived generativity in older adults' *Journal of Intergenerational Relationships* 12 pp 40–53. doi:10.1080/15350770.2014.870865.

Emihovich C (1995) 'Distancing passion: Narratives in social science' in Hatch J and Wisniewski R (eds) *Life History and Narrative*. Routledge Falmer: Abingdon, pp 37–48.

Fielding N (2006) 'Life history' in Jupp V (ed) *The Sage Dictionary of Social Research Methods*. Sage: London, pp 159–161.

Flyvbjerg B (2006) 'Five misunderstandings about case-study research' *Qualitative Inquiry* 12 (2) pp 219–245. doi:10.1177%2F1077800405284363.

Foster D (2016) 'It's not just London that has a housing crisis' *The Guardian*. Available at https://www.theguardian.com/housing-network/2016/jun/17/london-housing-crisis-homelessness-poverty-eviction. Accessed 8 July 2017.

Foucault M (1972) *The Archaeology of Knowledge*. Routledge: London.

Foucault M (1976) 'Power/knowledge' in Delanty G and Strydom P (eds) (2003) *Philosophies of Social Science: The Classic and Contemporary Readings*. Open University Press: Philadelphia, PA, pp 346–353.

Fraser I (2015) 'Mayor Joe Anderson hints at Giants return to Liverpool in 2016' *Liverpool Echo*, 21 January 2015. Available at http://www.liverpoolecho.co.uk/news/liverpool-news/mayor-joe-anderson-hints-giants-8424325. Accessed 21 May 2017.

Furmedge P (2008) 'The regeneration professionals' in Allt N (ed) *The Culture of Capital*. Liverpool University Press: Liverpool, pp 82–97.

Gaillard B and Rodwell D (2015) 'A failure of process? Comprehending the issues fostering heritage conflict in Dresden Elbe Valley and Liverpool – Maritime Mercantile City world heritage sites' *The Historic Environment: Policy and Practice* 6 (1) pp 16–40. doi:10.1179/1756750515z.

Gardner P (2006) 'Oral history' in Jupp V (ed) *The Sage Dictionary of Social Research Methods*. Sage: London, pp 206–208.

Getting By? (2015) *Getting By? A Year in the Life of 30 Working Families in Liverpool*. Available at www.gettingby.org.uk. Accessed 11 May 2015.

Gilbert A (2007) 'The return of the slum: Does language matter?' *International Journal of Urban and Regional Research* 31 (4) pp 697–713. doi:10.1111/j.1468-2427.2007.00754.x.

Glaser B and Strauss A (1967) *The Discovery of Grounded Theory: Strategies for Qualitative Research*. Aldine: New York, NY.

Goffman E (1996) *Behaviour in Public Places: Notes on the Social Organisation of Gatherings*. Free Press: New York, NY.

Harding J (2006) 'Grounded theory' in Jupp V (ed) *The Sage Dictionary of Social Research Methods*. Sage: London, pp 131–132.

Harvey L (1990) *Critical Social Research*. Unwin Hyman: London.

Hayes G (1987) *Past Trends and Future Prospects: Urban Change in Liverpool 1961–2001*. Liverpool City Council: Liverpool.

Hayward G (1977) 'Psychological concepts of "home"' *HUD Challenge* 8 (2) pp 10–13.

Holton J (2007) 'The coding process and its challenges' in Charmaz K and Bryant A (eds) *The Sage Handbook of Grounded Theory*. Sage: London, pp 266–289. doi:10.4135/9781848607941.n13.

Howarth K (1998) *Oral History: A Handbook*. Sutton Publishing: Stroud.

Janesick V (2007) 'Oral history as a social justice project: Issues for the qualitative researcher' *The Qualitative Report* 12 (1) pp 111–121.

Kearon T and Leach R (2000) 'Invasion of the "Body Snatchers": Burglary reconsidered' *Theoretical Criminology* 4 (4) pp 451–472. doi:10.1177%2F1362480600004004003.

Kelly L, Burton S and Regan L (1994) 'Researching women's lives or studying women's oppression? Reflections on what constitutes feminist research' in Maynard M and

Purvis J (eds) *Researching Women's Lives from a Feminist Perspective*. Taylor and Francis: Abingdon, pp 27–48.

Kelly S (2015) '11 regional British stereotypes as portrayed on screen' *The Independent*, 20 June 2015. Available at http://www.independent.co.uk/arts-entertainment/tv/features/11-regional-british-stereotypes-as-portrayed-on-screen-10331867.html. Accessed 7 July 2017.

Kidd S and Evans D (2011) 'Home is where you draw strength and rest: The meanings of home for houseless young people' *Youth & Society* 43 pp 752–773. doi:10.1177%2F0044118X10374018.

Kinsella C (2011) 'Welfare, exclusion and rough sleeping in Liverpool' *International Journal of Sociology and Social Policy* 31 (5/6) pp 240–252. doi:10.1108/01443331111141246.

Kirby D (2016; updated 2019) 'Liverpool at risk of losing World Heritage Site status' *iNews*, 11 October 2016; updated 6 September 2019. Available at https://inews.co.uk/news/uk/liverpool-risk-losing-world-heritage-site-status-533871. Accessed 5 May 2020.

Koumenta M and Williams M (2019) 'An anatomy of zero-hour contracts in the UK' *Industrial Relations Journal* 50 (1) pp 20–40. doi:10.1111/irj.12233.

Leeson P (1970) *Us and Them*. Documentary. Available at http://www.peterleeson.co.uk/liverpool/. Accessed 22 June 2016.

Lempert L (2007) 'Asking questions of the data: memo writing in the grounded theory tradition' in Charmaz K and Bryant A (eds) *The Sage Handbook of Grounded Theory*. Sage: London, pp 243–264. doi:10.4236/ib.2011.33031.

Lewicka M (2005) 'Ways to make people active: The role of place attachment, cultural capital, and neighbourhood ties' *Journal of Environmental Psychology* 25 pp 381–395. doi:10.1016/j.jenvp.2005.10.004.

Lewicka M (2010) 'What makes a neighbourhood different from home and city? Effects of place scale on place attachment' *Journal of Environmental Psychology* 30 pp 35–51. doi:10.1016/j.jenvp.2009.05.004.

Liverpool City Council (2016) *Housing Strategies: Empty Homes Strategy*. Available at http://liverpool.gov.uk/council/strategies-plans-and-policies/housing/housing-strategy/empty-homes-strategy/. Accessed 8 July 2017.

Liverpool Waters (2019) *Huge Vote of Public Support for New Everton Stadium and Legacy Plans*. Available at https://liverpoolwaters.co.uk/public-support-for-everton-stadium-plans/. Accessed 28 April 2020.

Longman D (2016) *Liverpool in the Headlines*. Amberley: Stroud.

Mac an Ghaill M and Haywood C (2011) '"Nothing to write home about": Troubling concepts of home, racialization and self in theories of Irish male (e)migration' *Cultural Sociology* 5 (3) pp 385–402. doi:10.1177%2F1749975510378196.

McLean R, Robinson G and Densley J (2020) *County Lines, Criminal Networks and Evolving Drug Markets*. Springer: Switzerland. doi:10.1007/978-3-030-33362-1.

Mallett S (2004) 'Understanding home: A critical review of the literature' *Sociological Review* 52 (1) pp 62–89. doi:10.1111/j.1467-954X.2004.00442.x.

Manturuk K, Lindblad M and Quercia R (2010) 'Friends and neighbours: Home ownership and social capital among low- to moderate-income families' *Journal of Urban Affairs* 32 (4) pp 471–488. doi:10.1111/j.1467-9906.2010.00494.x.

Massey D (1994a) 'Double articulation: A place in the world' in Bammer A (ed) *Displacements: Cultural Identities in Question*. Indiana University Press: Bloomington, pp 110–121.

Massey D (1994b) *Space, Place and Gender*. University of Minnesota Press: Minneapolis, MN.

Ministry of Housing, Communities and Local Government (2019) *The English Indices of Multiple Deprivation.* Available at https://assets.publishing.service.gov.uk/government/uploads/system/uploads/attachment_data/file/835115/IoD2019_Statistical_Release.pdf. Accessed 29 April 2020.

Morley D (2000) *Home Territories: Media, Mobility and Identity.* Routledge: Abingdon.

Muchnick D (1970) *Urban Renewal in Liverpool.* The Social Administration Research Trust: Birkenhead.

Murphy L (2016) 'Revealed: The futuristic blueprint for a Merseyside "tube map"' *Liverpool Echo,* 7 May 2016. Available at http://www.liverpoolecho.co.uk/news/liverpool-news/merseyside-liverpool-tube-map-2080-11297427. Accessed 5 July 2017.

North West Research (2019) *Tourism Data Summary 2019.* Available at https://www.liverpoollep.org/wp-content/uploads/2019/08/Tourism-Data-Summary-August-2019.pdf. Accessed 28 April 2020.

Parr J (2017) 'The Northern powerhouse: A commentary' *Regional Studies* 51 (3) pp 490–500. doi:10.1080/00343404.2016.1247951.

Parry J (2016) 'Liverpool in housing crisis – With only one home built for every eight new arrivals' *Liverpool Echo,* 10 September 2016. Available at http://www.liverpoolecho.co.uk/news/liverpool-news/liverpool-housing-crisis-only-one-11866028. Accessed 2 June 2017.

Paton K and Cooper V (2017) 'Domicide, eviction and reposession' in Cooper V and Whyte D (eds) *The Violence of Austerity.* Pluto Press: London.

Platt L (2016) 'Dealing with the myths: Injurious speech and negative interpellation in the construction of tourism places' in Andrews H (ed) *Tourism and Violence.* Ashgate: Abingdon, pp 69–83.

Polkinghorne D (1995) 'Narrative configuration in qualitative analysis' in Hatch J and Wisniewski R (eds) *Life History and Narrative.* Routledge Falmer: Abingdon, pp 5–23.

Pooley C and Irish S (1984) *The Development of Corporation Housing in Liverpool 1869–1945.* Centre for North West Regional Studies: Lancaster.

Reed-Danahay D (2006) 'Autoethnography' in Jupp V (ed) *The Sage Dictionary of Social Research Methods.* Sage: London, pp 15–16.

Robinson G, McLean R and Densley J (2019) 'Working county lines: Child criminal exploitation and illicit drug dealing in Glasgow and Merseyside' *International Journal of Offender Therapy and Comparative Criminology* 63 (5) pp 694–791. doi:10.1177/0306624X18806742.

Rose A (2001) 'Going deep: Intersecting the self as researcher and researched' in Cole A and Knowles J (eds) *Lives in Context: The Art of Life History Research.* Alta Mira: Oxford, pp 170–176.

Roth W (2009) 'Auto/ethnography and the question of ethics' *Forum: Qualitative Social Research* 10 (1) pp 1–10.

Rybczynski W (1986) *Home: A Short History of an Idea.* Pocket Books: London.

Rykwert J (1991) 'House and home' *Social Research* 58 (1) pp 51–62.

Sadd D (2010) 'What is event-led regeneration? Are we confusing terminology or will London 2012 be the first games to truly benefit the local existing population?' *Event Management* 13 pp 265–275. doi:10.3727/152599510X12621081189112.

Scanlon K, Lunde J and Whitehead C (2011) 'Responding to the housing and financial crises: Mortgage lending, mortgage products and government policies' *International Journal of Housing Policy* 11 (1) pp 23–49. doi:10.1080/14616718.2011.548585.

Somerville P (1992) 'Homelessness and the meaning of home: rooflessness or rootlessness?' *International Journal of Urban and Regional Research* 16 pp 529–539.

Snider L (2018) 'Enabling exploitation: Law in the gig economy' *Critical Criminology* 26 pp 563–577. doi:10.1007/s10612-018-9416-9.

Standing G (2011) *The Precariat: The New Dangerous Class*. Bloomsbury: London.

Taylor K (2010) 'It's time to drop the negative stereotyping of Liverpool' *The Guardian*, 11 November 2010. Available at https://www.theguardian.com/commentisfree/2010/nov/11/drop-negative-stereotyping-liverpool. Accessed 7th July 2017.

Thomas J (2019) '5 murders and 80 shootings: Merseyside's war on gun crime' *Liverpool Echo*, 14 April 2019. Available at https://www.liverpoolecho.co.uk/news/liverpool-news/blood-tears-tributes-murders-rise-16107787. Accessed 5 May 2020.

Throgmorton J (2003) 'Planning as persuasive storytelling in a global scale web of relationships' *Planning Theory* 2 (2) pp 125–151. doi:10.1177%2F14730952030022003.

Topping P and Smith G (1977) *Government against Poverty? Liverpool Community Development Project 1970–75*. Social Evaluation Unit: Oxford.

Tuan Y (1977) *Space and place: The perspective of experience*. University of Minnesota Press: Minneapolis.

UNESCO (nd) *About World Heritage*. Available at http://whc.unesco.org/en/about/. Accessed 1 June 2016.

Urquhart C (2013) *Grounded theory for qualitative research: A practical guide*. Sage: London.

Waghorn K (2009) 'Home invasion' *Home Cultures* 6 (3) pp 261–286. doi:10.2752/174063109x12462745321507.

Wilson H and Womersley J (1977) *Change or Decay: Final Report of the Liverpool Inner Area Study*. HMSO: London.

Windsong E (2010) 'There is no place like home: Complexities in exploring home and place attachment' *The Social Science Journal* 47 pp 205–214. doi:10.1016/j.soscij.2009.06.009.

Wood D and Beck R (1994) *Home Rules*. Johns Hopkins University Press: Baltimore, MD.

Woodside A (2010) *Case Study Research: Theory, Methods and Practice*. Emerald: Bingley.

Wright G (1991) 'Prescribing the model home' *Social Research* 58 (1) pp 213–225.

Zeller N (1995) 'Narrative strategies for case reports' in Hatch J and Wisniewski R (eds) *Life History and Narrative*. Routledge Falmer: Abingdon, pp 75–85.

2 The meaning(s) of home

Introduction

'Home' simultaneously manages to be the most simple and the most difficult thing to define (Porteous and Smith, 2001) – it is at once familiar, so familiar in fact that it does not need explanation, and elusive – difficult to pin down and conceptualise. Various academics have attempted to theorise the concept, but none have really been able to isolate the precise, special quality of home which renders it different from any other space or place. Simultaneously, home is by necessity a multi-faceted and multi-level concept, taking many forms with varying degrees of importance contingent on a range of variables. Thus, home in some senses is enigmatic; Fox (2002, p590) employs the term "the *x* factor" (emphasis in original) to label the aspects of home which make it both special and difficult to categorise.

The purpose of this chapter is to provide a theoretical explanation and academic 'reading' of home in its various forms. Starting with the etymology of the term, I explore how home emerged as a concept and became standardised alongside other pillars of the capitalist system, including the nuclear family, patriotism, and commodification. Status in relation to home is explored, as is its impact on life chances, whilst the related concepts of space, place, mobility, and stasis are considered as mechanisms to facilitate both the positive and the negative aspects of home. The chapter closes with a discussion on homelessness as constructed as the binary opposite to home, and the experience of loss of home.

The etymology and development of the terms 'house' and 'home'

In the search for the sociological and criminological meaning of the concept of home, a possible starting point may be the root and development of the word 'home'; its associated concept, 'house'; and the related terms 'domicile' and 'dwelling'. Words used in other languages that are translated as the English word 'home' are not necessarily an accurate equivalent and are reflective of the various socio-geographic meanings, cultural meanings, and emphases attached to the concept.

There are several Latin words that are identifiable as meaning 'home' in one way or another (Rykwert, 1991). *Aedes*[1] refers to a structure that has been built, whilst *mansion* refers to "a place of rest" (ibid., p63). *Domus* in Latin is the closest direct translation of home, and the noun *casa* is "a hut or cottage, something humble" (ibid.). Although all of these words remain conceptually relevant to contemporary understandings of home, only three are directly recognisable within the Romance languages today. *Casa* is the Spanish word in common use for home, and has, over a long period, evolved into the French word *chez* to indicate home in the sense of 'my place' or 'your place' (*chez moi; chez toi*) (ibid.). The Latin word *domus* continues to have a place in the Romance languages in terms of indicating home, as well as "domesticity, household effects, homeliness – even peace" (ibid.). Crucially, it continues to embody the "symbolic power of 'domain' and 'domination'" (Wright, 1991, p214). Inside the home, the *focus* – taken contemporarily in English to mean a point to concentrate on, or an item to pay most attention to – was the "'hearth' or 'fireplace', which even in classical times served as a metonymy for 'household' or 'family'" (Hollander, 1991, p42).

Whilst the ancient Greeks also used the word *domus* to connote 'house', a different word, *oikos*, connotes 'home' (Rykwert, 1991). The two concepts were distinct enough to warrant the development of a compound word – *oikodomein* – indicating a structure that has been built with the purpose of "sheltering the home" (ibid., p52). *Oikos* is the root term of several words (ibid.), perhaps most notably ecology, the "household of nature" (Anker, 2003; cited in Blunt, 2005, p511) – pointing perhaps to the idea that the natural environment is the ultimate home (ibid.). The Greek term for rural or rustic settlement – *khomi* – has developed from the ancient Indo-European word *kei* which means 'to settle' in the literal sense of lying down, resting (Rykwert, 1991), and can also connote "something dear or beloved" (Hollander, 1991, p44). Interestingly, *kei* also provides the root of the word 'cemetery', indicative of the notion of death as the ultimate home in that it is the ultimate destination (Porteous and Smith, 2001).

Modern German provides us with a helpful insight into how 'home' has come to have conceptually different meanings. In German, a micro/macro distinction can be drawn between *heim* – the private home where one lives, in the sense of a dwelling, shared with a small number of people or perhaps no one – and *heimat* – the public home where one lives, in the sense of a homeland, shared with many others (Morley, 2000; Moore, 2000). Words related to *heim* and *heimat* provide further clues as to how ideas about what home means have evolved – *geheim*, for example, means 'secret' – something so private, so 'of the home', that it is concealed and esoteric (Hollander, 1991). *Unheimlich*, whilst appearing to translate directly to 'unhomely', translates more correctly to 'uncanny', 'creepy', and 'incredible' – something so removed from 'home' that it is both unbelievable and disturbing (Kearon and Leach, 2000). Meanwhile, similar concepts and ideas are prioritised in Irish Gaelic. Mac an Ghaill and Haywood (2011) note that the Irish word *baile* refers to home both in the sense of dwelling or place of shelter, and also one's hometown, as a place of belonging evocative of feelings of love and being loved (Fox, 2002).

Wright (1991, p214) contends that

> the very word 'home' is unique to the English language, after all, albeit with potent similarities to the German Heim, offering protection and familiarity, and allusions to the walls of the Old English *hus*, where goods were safely stored and husbanded (emphasis in original).

In other words, the English word 'home' encapsulates elements of both the sentiments of *heim* and the practicalities of *hus*. However, the word itself derives from the old English *ham*, meaning village or estate (Mallett, 2004), evident today as a suffix in place names. Coolen and Meesters (2012) however note that another word in common usage in English, 'dwelling', is used as a verb as well as a noun – 'to dwell' as in to linger or to stay a while, also to contemplate or consider at length – as Kearon and Leach (2000, p458) put it, "dwelling is to be taken literally". In this sense, the 'home', or 'dwelling', is intimately connected to existence itself in that it is reflective of the experience of life and how it can be processed and 'thought through' in a safe, comfortable environment (Cuba and Hummon, 1993; Smith, 1994b; Cristoforetti et al., 2011). Crucially, the culture of the home influences, and is influenced by, the thought or 'dwelling' processes of those who 'dwell' within it (Blunt, 2005), indicated by Kearon and Leach who note that "cultural theory is rediscovering 'dwelling' as a rich and ambiguous category within the social that denotes the experience of domestic space rather than its status as the bearer of structure" (2000, p458).

Just as it is important to understand 'dwelling' as both a verb and a noun, it is also necessary to recognise that 'house' and 'home' operate in the same way. Whilst the word 'house' is clearly a noun, those who reside within a house are 'housed' there and may have been 'housed' by an external body like, for example, a local authority. In this sense, housing is something that can be 'done' to you; a process that you can be subjected to in a passive manner. Similarly, the word 'home' is clearly a noun but, as Mallett (2004) argues, 'home' is also something that people 'do', or have 'done' for them – it is a process as much as it is a fixed place. As Ginsburg (1998, cited in Mallett, 2004, p83 puts it, "our residence is *where* we live, but our home is *how* we live" (emphasis added).

The conceptual difference between house and home, then, revolves around the notion of a home being something much more than just a house, of having much more significance and value attached to it, a deeper meaning than mere bricks and mortar (Dovey, 1985; Kidd and Evans, 2011; Windsong, 2010). In some senses, the difference between house and home is an issue of morality, as indicated by the common practice of shortening the word 'whorehouse' to 'house' leading to the use in language of 'home' to denote respectability (Hollander, 1991). Perhaps though the difference between house and home is elucidated most clearly by exploring the misuse of the term 'home' in the housing market (Mallett, 2004; Coolen and Meesters, 2012), as illustrated by Hollander (1991, p41):

> The common – and, unlike many common expressions, vulgar – use of 'home' as a euphemism for 'house' is by and large the linguistic waste product of the American real estate industry…a student of mine recalls a Chinese émigré academic who had taught her in college reading aloud in disbelief an ordinary sign: "*Homes for Sale?*" How can you buy or sell a home? Home is (and he groped for the formulation) … memories.

Arguably, this is where our academic, conceptual problems with defining and understanding 'home' begin – we link it too much to the 'house' and all its associated meanings.[2] Mallett (2004, p66) contends that

> governments of advanced capitalist countries have actively promoted the conflation of house, home and family as part of a broader ideological agenda aimed at increasing economic efficiency and growth. These governments have attempted to shift the burden of responsibility for citizen's welfare away from the state and its institutions on to the home and nuclear family.

This notion of the concept of home being reshaped and distorted to capitalist ends will inform and frame discussion throughout the remainder of the chapter.

Meanings of home

As we have begun to discover in the previous sections, there is no single, clear, comprehensive understanding of home. Rather, home is an ambiguous and amorphous concept which can be observed from different angles and can be subjected to a range of academic analyses (Dovey, 1985; Douglas, 1991; Massey, 1994; Mallett, 2004). In this section, I explore the notion of home from three separate points of departure. First, I look at the idea of home in relation to place, starting at the micro level and working through to the macro level. Second, I consider home from a phenomenological perspective and conceptualise home in relation to feelings, emotions, and experience. Third, I look at home in relation to the passage of time, linking the concept to individual development, the movement from childhood to adulthood within the family setting, and the development of memories. Although these starting points are presented in this introductory paragraph as distinct from each other, it must be noted that they are related and interwoven and are in themselves reflective of the complexity of the nature and meaning of home.

Home as place: Concentric circles

When examining the literature on the concept of home, a recurrent theme that is apparent is the notion of 'levels' of home (Hobsbawm, 1991; Hollander, 1991; Case, 1996; Porteous and Smith, 2001; Mallett, 2004; Lewicka, 2010; Gillsjo and

Schwartz-Barcott, 2011; Kidd and Evans, 2011). If we attempt to conceptualise home as a physical place, we must within that recognise that there are differing degrees of home 'status' attached to places – ranging from the micro level (one's bed, for example) through to the macro level (nation or homeland) – and that differing degrees of importance will be attached to each level depending on individual interpretation. In this sense, it is possible to explore the concept of home as place as if home were a series of Russian dolls tucked inside one another or, perhaps more accurately, a series of concentric circles radiating out from a central starting point (Hayward, 1977).

For Hayward, the nucleus of these concentric circles is not the physical structure of the residence itself; rather, it is the bed – the ultimate representation of grounding and comfort (Case, 1996).[3] The second 'layer' of home in this analysis is the bedroom, or the place within the residence where individuals can, away from those who share the residence, enjoy solitude (Tamm, 1991), hide their secrets (Porteous and Smith, 2001), and experience a sensation of "inviolable space" (Cristoforetti et al., 2011, p229). The bedroom can act as a symbol of the togetherness and intimacy of a relationship (ibid.) or, equally, as *un lit a soi*, a symbol of independence and freedom (de Beauvoir, 1997). However, within the physical structure of the residence there is a rival for the title of the nucleus of the home – the hearth (Somerville, 1992). Rykwert (1991) argues that the hearth – the place of fire within the home – is the ultimate focal point due to its role in providing heat, light, cooked food, and a sense of safety from predators. Rykwert notes of the hearth that "eating is usually done close to it, so that eating and sleeping together have come to define the household" (p51) – in this sense, it is the marriage of the bed and the hearth that typifies the home.

Just as spaces *within* the physical structure of one's residence can be representative of the essence of home, places *outside* the residence can also embody the meaning of home (Cuba and Hummon, 1993; Morley, 2000; Charleston, 2009).[4] Continuing the analogy of concentric rings, home *outside* of the immediate residence starts with the concept of neighbourhood or, for some, community (Meegan and Mitchell, 2001; Porteous and Smith, 2001; Lewicka, 2010; Manturuk et al., 2010). Described by Lewicka (2010, p36) as "the midpoint on the scale of place continuum", the concept of neighbourhood is more meaningful to some than the concept of nation – in fact, as Mac an Ghaill and Haywood (2011, p389, drawing on Brennan, 1990) point out, the Latin term *natio* connotes both 'nation' and "something more ancient and nebulous … a local community, domicile, family, condition of belonging".[5]

Neighbourhood

Something of an ambiguous concept, a neighbourhood area can be identified by the nature of its built environment, by the characteristics and qualities of those who reside within it, or perhaps by its historical and cultural significance (Lewicka, 2010), and can be interpreted differently by people according to size and space – "from the smallest block scale (equivalent to an area where children are allowed to play without supervision) to the whole sector of a city" (ibid., p37). However, the significance of neighbourhood extends beyond the concepts of space and place. Meegan and Mitchell (2001, p2172) understand a neighbourhood to be

> A key living space through which people get access to material and social resources, across which they pass to reach other opportunities and which symbolises aspects of the identity of those living there, to themselves and to outsiders.

Similarly, Lewicka (2005, p383) cites Ossowski's term "little fatherland" to indicate that place attachment operates more effectively in terms of locale than nation – "affective, cultural distinctions rather than territorial boundaries ... based on direct experience of locality" (Galbraith, 1997, p122). Similarly, Porteous and Smith (2001, p232) point to the Uzbek concept of the *mahalla*;[6] "a distinct and identifiable social neighbourhood, district or urban ward ... . Each *mahalla* is its own world, with its own rules. Uzbeks cannot imagine a person without a *mahalla*". A person's level of identification with or attachment to a neighbourhood can be dependent on length of residence within the neighbourhood (Cuba and Hummon, 1993), local levels of socio-economic deprivation (Jupp, 2013), the extent of social participation in the neighbourhood (Lewicka, 2005), and even whether they own the property within which they live (Bone and O'Reilly, 2010).[7]

For some, neighbourhood is synonymous with community, yet for others each should be treated as distinct concepts. For Lewicka (2005), a neighbourhood only develops a true sense of community where there are high levels of civic engagement. Meanwhile Smith (1994b) and Porteous and Smith (2001) agree that the destruction of neighbourhoods via, for example, 'slum' clearance programmes, inevitably leads to the destruction of community. However, community does not necessarily need to be attached to a particular area or locale. Bone and O'Reilly (2010, p247) talk of a "postmodern conception of community, one that unties its relationship to place" – this can be applied in terms of, for example, an interest community (Fischer, 2001), or indeed a 'virtual' (online) community (Morley, 2000; Carlen and Jobring, 2005). In any event, a sense of community can offer more of a sense of home to people than the physical structure within which they live (Waghorn, 2009).

Arguably though, for some people the concept of home city is more meaningful than neighbourhood (Lalli, 1992; Felonneau, 2004; Ng et al., 2005). The collective identity, reputations, and stereotypes associated with cities and indeed regions (Cuba and Hummon, 1993; Porteous and Smith, 2001) can speak of home more to an individual in that they signify certain qualities and characteristics, real or imagined, which are unique to that particular area (Kidd and Evans, 2011). In this sense, it is the specific, singular qualities of a given city or region which mark it out as unique as being 'home', as identified by a respondent of Kidd and Evans (2011, p764):

> Home is not some place where you live. Home is the place where you feel accepted. A place where you feel you belong. NYC [New York] itself is a place that I consider my home. It's a place where I come and I feel like people understand me. It's familiarity. It's a feeling you get.

Homeland

Then of course there is the notion of 'homeland' – nation, or country of origin as home (Case, 1996; Moore, 2000; Morley, 2000; Porteous and Smith, 2001; Waghorn, 2009). Porteous and Smith (2001, p42) cite Berger (1990) who "suggests that the word 'home' has been taken over by two kinds of moralists: the defenders of domestic morality and property (including women), and those who defend the notion of homeland". Thus, in contrast to the concept of *heim, heimat* – the national home – is more meaningful to many, and more evocative of home, than the bricks and mortar within which they live (Morley, 2000). Many see their country of birth as their true home (Moore, 2000; Waghorn, 2009), tied up with notions of patriotism and an almost child-parent love for the 'motherland' or 'fatherland' (Morley, 2000). In this sense one's homeland is the centre of one's world (Tuan, 1977). However, Mallett (2004, p65) notes that the development and fostering of love for one's homeland is something of a hegemonic scam, in that, as capitalism advanced, "the concept of homeland was appropriated by the ruling classes to promote a form of nationalism and patriotism aimed at protecting their land, wealth and power". In this sense, homeland or *heimat* is nothing more than a social construction (Hollander, 1991; Morley, 2000). Further, Porteous and Smith (2001, p240) argue that the concept of homeland as supportive of wealth accumulation will outlive its usefulness as

> geopolitically, nation states will remain in name only, since the development of supranational organisations and free-trade associations will end the ability of states to protect their citizens.

Finally, it cannot be ignored that, for many, a true sense of where home *is* only develops when one has spent an amount of time *away* from home. In the same way that people develop their understanding of what home means to them by being away from it (Case, 1996), those who migrate from their home country, or are exiled from it, acquire distinctive and complex understandings of the home they have left behind (Hobsbawm, 1991; Hollander, 1991; Rykwert, 1991; Morley, 2000; Porteous and Smith, 2001; Weiman and Wenju, 2007; Mac an Ghaill and Haywood, 2011). Notions of *heimat* are mirrored by notions of *fremde* – a 'host' nation characterised by loneliness, remoteness, strangeness, and ill-acquaintance (Morley, 2000).

For some who develop a sense of *fremde* in a foreign land, a sense of *heimweh* naturally accompanies it. More than simply homesickness or *mal du pays*, *heimweh* encapsulates a nostalgic longing for a lost feeling:

> *Heimweh* [is] considered as a mode of something like melancholia. We designate by it not a literal "homesickness" but a strange and not quite legitimate expression thereof, a longing for a time and not a place.
>
> (Hollander, 1991, p37, emphasis in original)

For those afflicted by *heimweh*, 'going home' can only ever mean returning to one's *heimat* (ibid.), and *heimweh* can never really be alleviated:

> Just so, the migrant's adopted home is never home, but the migrant is too changed to be welcome in her own country. Only in dreams will she see the skies of home.
>
> (Greer, 1993, cited in Porteous and Smith, 2001, p57)

Some attempt to overcome feelings of *heimweh* through creating an atmosphere of *heimat* within a climate of *fremde*, by associating with those with the same national heritage and searching for authenticity in a shared cultural milieu (Porteous and Smith, 2001). However, the goal of turning *fremde* into *heimat* is difficult to achieve with accuracy. Morley (2000) points to the Southall area of East London as an example of a community so entrenched in its need to recreate the culture of the Indian cities that they left behind that they have built a way of life stuck in time and bearing little resemblance to contemporary India.[8] Oftentimes, as people frequently are compelled to leave their home nation for economic reasons, links to home nations are financial – Mac an Ghaill and Haywood (2011) note that money earned in wages being 'sent back home' is as much an act of duty to the home nation state as it is an act of love and loyalty to one's family. The need for economic survival can however seemingly undermine romantic and sentimental conceptions of home – as a Turkish migrant cited by Morley (2000, p44) sums it up, "home is wherever you have a job".

The idea of 'roots' is a crucial concept in terms of place attachment, be it in relation to neighbourhood, city, region, or nation (Relph, 1976). Experiencing a sense of rootedness is considered to be a human need which fulfils both spiritual and psychological requirements (Windsong, 2010) and affords a sense of belonging (Charleston, 2009). An individual's sensation of 'roots' can be reflective of place of birth (Hobsbawm, 1991), ethnic background (ibid.; Weiman and Wenju, 2007; Oliver-Rotger, 2011; Tete, 2012) or, alternatively, cultural ideas of group affiliation (Charleston, 2009). In other senses, though, rootedness can be a much more abstract concept which has more to do with the "human soul" (Porteous and Smith, 2001, p52) and "a feeling rooted in the heart" (Weiman and Wenju, 2007, p184) than clearly identified place attachment (Hobsbawm, 1991). It is to the notion of home as a feeling rather than a place that we will turn next.

Home as experience: Feelings of home

Clearly, understandings of home are closely linked to sensations, emotions, and feelings. To 'feel at home' is to *be at home*, to be at complete comfort and ease with oneself (Seamon, 1979; Hollander, 1991). Cuba and Hummon (1993, p117) note that "expressions of at-homeness" are essentially an indicator of who a person really is. This inherent bond between notions of home and notions of true self – of being "really me" (ibid., p113) – are the starting point for an area of academic study of the home which focuses on the feelings, experiences, and processes associated with it. This approach to thinking about home can be broken down into three broad strands: the psychological, the sociological, and the phenomenological.

For some theorists, the feelings and experiences of home are best explored from a psychological perspective (Kearon and Leach, 2000; Moore, 2000) as the sensation of home is essentially a psychological phenomenon (Fox, 2005) which is inherently personal rather than social (Moore, 2000). Here it is psychological processes which render a home *more than* a house (Waghorn, 2009) – as Tomas and Dittmar (1995, p501) put it, "whereas a 'home' is understood to be psychologically meaningful, the 'house' assumes a neutral status as container". Psychological relationships with home can be reflective of one's life cycle (Porteous and Smith, 2001), as evidenced by Kearon and Leach (2000, p463), who liken our psychological relationship to home to a child's psychological relationship to a transitional object –

> any object adopted by the infant that comes to represent security in the absence of a parent – it does this by combining the qualities of a thing both loved and hated, both present and absent, both permanent and transitory.

In a similar way, the concept of 'holding' within children's psychology, whereby "the child needs to be held by the primary carer in order that it achieves a sense of continuity of experience" (ibid., p464), is analogous to the "bonds between people and *home* places" (Moore, 2000, p209, emphasis in original). However, as Tomas and Dittmar (1995) note, the physical structure of home is not strictly necessary for the psychological experience of home.

A further layer of understanding of feelings and experiences of home can be added through a sociological perspective (Hareven, 1991; Cuba and Hummon, 1993; Cristoforetti et al., 2011). Here the *meaning* (in the Meadian[9] sense) of home locates the home as central to the development of a sense of self and self-identity (ibid.). Our subjective relationship with the home (Porteous and Smith, 2001) goes beyond the psychological in that the home fulfils a range of emotional functions (Cuba and Hummon, 1993), which may be indicative of family relationships (Hareven, 1991), gender divisions (Cuba and Hummon, 1993), or social status (Cristoforetti et al., 2011). In this sense, homes become "meaning makers" (ibid., p226); indeed, as Coolen and Meesters (2012, p3) note, home, in itself, *is* meaning:

> House and dwelling [are] used to indicate the physical structure, while home [is] used for the relationships we experience within the physical structure and the meanings we attach to it. Given this distinction the use of the phrase *meaning of home* seems tautological, and a sign saying *home for sale* is meaningless (emphasis in original).

Thus, as well as a psychological significance, there is also a sociological significance to the relationship between people and their homes (Kearon and Leach, 2000).

The approach of taking feelings and experiences of home as a starting point can be further enhanced via phenomenology[10] (Dovey, 1985). For Dovey, the mere fact that home is at once an abstract, intangible concept, and the nucleus of a range of social interactions, means that it must be analysed from the point of view of experiences and processes rather than as a static inanimate object (ibid.). Thus, the home, in whatever form it takes, is the lived space necessary for "being in the world" (Heidegger, 1962; cited in Dovey, 1985, p34). From this phenomenological perspective, home is something that people 'do' – it is an active process (Mallett, 2004). Case (1996, p2) discusses transactionalism[11] and phenomenology as similar means of developing a conceptual framework for the understanding of home as experience, yet notes that only phenomenology "makes home the primary and central point from which the rest of the world is experienced and defined". However, as Somerville (1992, p530) points out, home does not have to be experienced to have meaning:

> Home is not just a matter of feelings and lived experience but also of cognition and intellectual construction: people may have a sense of home even though they have no experience or memory of it. ... We cannot know what home 'really' is outside of these ideological structures.

From each of these psychological, sociological, and phenomenological perspectives, the relationship between the home and the people who live within it is paramount. For many, the home is in some way symbolic of the self (Kearon and Leach, 2000; Fox, 2002) – be it as a 'showcase' of what a person feels themselves to be (Cristoforetti et al., 2011), a mirror (Cooper Marcus, 1995), a mechanism for self-expression (Kidd and Evans, 2011), an indicator of social rank (Cuba and Hummon, 1993) or a measure of self-esteem (Smith, 1994b) – as Despres (1991, p100) puts it, "after the body itself, the home is seen as the most powerful extension of the psyche". In this sense, there is an inherent link between the physical structure of the home and physical human body – the home becomes inscribed[12] (Skeggs, 2004) on the body to the point that the home becomes an extension of the physical body[13] (Kearon and Leach, 2000). A good illustration of the notion of home as body is the sense of physical violation that people feel after the home is 'invaded'[14] through burglary (ibid.; Hobsbawm, 1991). Further, Mulla (2008) demonstrates that, after sexual assault carried out within the home setting, criminal justice agents respond to the victim's body as if it were a scene of crime, prioritising the preservation of evidence to the point where the home and the assault become so enmeshed that "disturbed domestic tranquillity can be 'rewritten' or displaced onto the body" (p306). Thus, "home can literally be scraped from underneath the fingernails, or tweezed off the clothing" (p317).

Feelings and experiences of home are also crucial to the notion of place attachment (Cuba and Hummon, 1993; Lewicka, 2005, 2010; Windsong, 2010). Emotional ties to place and the sensation of "rootedness" (Charleston, 2009) are key elements in the development of self-identity (Smith, 1994b; Moore, 2000;

Cristoforetti et al., 2011) and group affiliation (Douglas, 1991; Charleston, 2009), and their importance is recognised in human rights legislation (Neild and Hopkins, 2013). Place attachment is commonly but erroneously[15] viewed as an inherent good (Lewicka, 2005) and linked to home ownership[16] (Manturuk et al., 2010); however Windsong (2010, p210), in her study of communes, found that people can experience a "deeper sense of ownership" of their home that goes beyond the idea of home as a financial asset. Further, Lewicka (2010, p42) notes that some studies have found that place attachment is enhanced when the physical characteristics of a place are perceived to render it beautiful – however, she cautions that "place attachment … is not the same as residence satisfaction or positive evaluation of residence place", citing post-Second World War Warsaw, and Chernobyl after the nuclear disaster, as examples whereby place attachment is by no means reliant on quality of life.

Feelings of security are strong indicators of place attachment (ibid.). Sensations of safety (Despres, 1991), sanctuary (Waghorn, 2009), haven (Hareven, 1991; Cristoforetti et al., 2011), control of territory supported by the law (Smith, 1994b), and refuge (Mallett, 2004) are closely associated with both experiences and perceptions of home, in spite of the fact that the home, for many, is a place of fear and danger rather than security[17] (Tyner, 2012) – an issue that I return to later. Hareven (1991, p38) locates the emergence of the concept of home as safe haven against the backdrop of developing urbanisation:

> The home began to be viewed as a Utopian community, as a retreat from the world – one that had to be consciously designed and perfectly managed. The idealisation of the home as a haven was a reaction to the anxiety provoked by rapid urbanization, resulting in the transformation of old neighbourhoods and the creation of new ones, the rapid influx of immigrants into urban areas, and the visible concentration of poverty in cities.

Accordingly, Rykwert (1991, p53) cites a statement from 17th-century judge Sir Edward Coke – "the house of everyman is to him as his castle and his fortresse, as well as his defence against injury and violence, as for his repose" – a sentiment contemporarily synopsised as 'an Englishman's home is his castle'. This position is indicative of two important points about the relationship between home and security; first, that military parlance[18] is commonly invoked to describe the protection that home offers from the outside world; second, that implicit in this is the assumption that "the outside world is a place to be feared" (Fox, 2002, p594), particularly for vulnerable groups such as the elderly (Cristoforetti et al., 2011). Specifically, the concept of ontological security is paramount here. Giddens (1990, p92) conceptualises ontological security as

> the confidence that most human beings have in the continuity of their self-identity and in the constancy of the surrounding social and material environments of action. A sense of the reliability of persons and things, so central to the notion of trust, is basic to feelings of ontological security.

Clearly, this can be related to the routine and familiarity of home, as evidenced by the disruption to perceived individual ontological security caused by home invasion:

> burglary is so emotionally and physically upsetting because it stimulates a sort of existential dread or 'bad faith': ontological security is immediately, instantaneously and lastingly revealed for the necessary fiction that it is.
>
> (Kearon and Leach, 2000, p466)

Thus, although feelings of safety and security are assumed features of home, they are not always present, but always desirable, particularly for those with a limited or idealised understanding of home. Tomas and Dittmar (1995, p453), in their study on the meaning of home to homeless women, conclude that

> paradoxically, it would appear that the meaning of 'home' cannot be separated from the experience of housing and yet survives in the absence of its most important functions (safety and security). ... The basic safety and security that housing was meant to provide, and homeless women had a limited experience of, appears to be all that these women are asking for in 'home'.

Home as memories: Family life and the passage of time

It would be erroneous not to recognise that there can be no understanding of the meaning of home without cognisance of the passage of time – as Dovey (1985, p35) puts it,

> Home as order is not only spatial orientation but also temporal orientation. Home is a kind of origin, we go 'back' home even when our arrival is in the future.

Porteous and Smith (2001, p39) note that "it is a common experience that a certain slanting of the light, or the smell of a spice from our past will bring home flooding back to us"; however, the relationship between home and time is twofold, in that it is reflective of both the passing of years (Coolen and Meesters, 2012) and the routines and rituals which make up the day-to-day time rhythms of the home (Douglas, 1991). The longing for home and the search for its "authenticity" (Massey, 1994) in connection to a particular time period operates both in terms of the past, for example, memories of a past that is happier than the present, and the future, for example, hopes for a happier home in times to come (Coolen and Meesters, 2012). Regarding the idea of longing for past home memories, Porteous and Smith (2001, p39) argue that "home is currently vying with nature as the post-Romantic or postmodern replacement for God. Nostalgia is indeed rampant". Thus, nostalgia for home is an almost universal human drive for those with experience of a 'happy' home (ibid.). Further, Douglas (1991, p294) notes that memory is crucial to the planning and organisation of home:

> Response to the memory of severe winters is translated into capacity for storage, storm windows, and extra blankets; holding the memory of summer droughts, the home responds by shade-giving roofs and water-tanks. Those are annual rhythms, but there are longer cycles, as testified by the standard pair of coffin stools always ready for the funeral wake in East Anglian houses.

Memories of home for many people revolve around the 'framework' of home imprinted on us by our childhood homes (Hayward, 1977; Douglas, 1991; Mallett, 2004) – the home of our childhood is our 'real' home (Brink, 1995). Implicit in this is the clear association between childhood home and family life (Wright, 1991; Fox, 2002, 2005; Coolen and Meesters, 2012). The "social and cultural unit" (Fox, 2002, p590) of the family home is constructed as being characterised by affection, love, and sentimentality (ibid.), associated with familiarity and ritual, for example, returning to the family home for special occasions (Morley, 2000), and family traditions observed (Douglas, 1991). In this sense, the home is the arena for conducting family life (Coolen and Meesters, 2012). Mallett (2004, p74) notes that "without the family a home is 'only a house'", whilst providing a family 'home' is a strong parental drive (Wright, 1991). For Fox (2005, p43) "it is the presence of children that makes a house into a home" – a claim borne out by case law which prioritises the family home over home *per se* (ibid.). However, it must be noted that the nuclear family household is essentially a phenomenon of industrialised society – Hareven (1991) demonstrates that in pre-industrial times 'family' was more commonly taken to mean 'household' rather than blood relations and work and family relationships, like work and rest spaces, were merged rather than separated. Therefore, the 'traditional' conception of the family home is a relatively modern construct (Wright, 1991). Further, as highlighted by Despres (1991), research on the family home has focused in the main on White, middle-class, nuclear families, thereby discounting versions of the family home which do not fulfil rigid, 'traditional' criteria, for example, single parent households (Coolen and Meesters, 2012).

Whilst a 'good' childhood home is central to 'good' child psychological development (Campbell and Parcel, 2010), Douglas (1991, p287) suggests that the childhood home is remembered with a "mixture of nostalgia and resistance", in that it is the site simultaneously of enrichment and suffocation, nurturing and control, support and suppression: "this is how the home works. Even its most altruistic and successful versions exert a tyrannous control over mind and body" (ibid., p303). Children and young people frequently experience home as an oppressive environment[19] (Porteous and Smith, 2001) characterised by rules, as synopsised by Morley (2000, p20) –

> If the childhood home gives initial shape to all later memory, this is perhaps because it is through (literally) learning to live ('behave') in the home that children learn the 'habitus' of their culture.

Wood and Beck (1994, p1) précis the relationship between children and home thus: "What is home for a child but a field of rules?" Children and young people frequently seek to escape this surveillance environment by looking for *privacy* in *public* space – Warpole (1995; cited in Morley, 2000, p57) highlights that

> many people, particularly young people, went to the park to escape from the overly *public* nature of domestic life … to seek an element of personal privacy on a park bench … as a space that was seen as more respectful and trustworthy than domestic space (emphasis in original).

Whilst "appearance can lead us to expect that family life should work out because the setting looks right" (Wright, 1991, p217), in reality, family home life, and negative aspects of it, frequently contribute to youth homelessness (Hagan and McCarthy, 1997; Kidd and Evans, 2011). It is the relationship between childhood, rules, and surveillance that renders a person's tenure in the family home necessarily limited; as children grow into adults, their parents have less and less entitlement to subject them to regulation (Mallett, 2004). The Western expectation is that, in adulthood, individuals will create their own home and, accordingly, their own rules (Wood and Beck, 1994). Increasingly, though, the autonomy associated with the relationship between adulthood and home is diminishing at both ends of the age spectrum, with younger adults less and less able to afford to buy or even rent their own homes due to prohibitive housing markets (Bone and O'Reilly, 2010), and expectations on older adults to make sense of a new 'home' environment in a nursing home setting (Woodhouse, 2006) where rules are established, not by themselves, but by 'powerful' health care professionals (McCloskey, 2011).

As we have seen throughout this section, home cannot have one single meaning – its meaning is contingent on time, place, experience, and interpretation. The ways in which experiences of home are governed by both determining contexts and broader social, cultural, and economic processes is considered in the next section.

Home as space, place, mobility, and stasis

For some people, home is clearly an aspect of space, filled with various life-enhancing elements, including privacy, control, freedom, safety, comfort, and leisure. For others, however, it is effectively a prison; a location where one is 'placed', situated against one's will and characterised by domination, subordination, dependence, and captivity. Similarly, mobility and stasis, whilst expressed as polar opposites, can and do manifest themselves both positively and negatively. This section considers how understandings of these concepts relate to home and shape the way it is experienced on several different levels.

Separating 'place' from 'space'

Withers (2009, p638) contends that in recent years, academia has been subject to what he terms a "spatial turn" in that scholars have turned their attention to 'space' and 'place' and attempted to interrogate and, subsequently, theorise them.[20] Some commentators seem to continue to view 'place' and 'space' as interchangeable terms (see e.g., Porteous and Smith, 2001; Gillsjo and Schwartz-Barcott, 2011), and the words are frequently treated as such in common parlance. Equally, however, there are inherent, subtle differences in interpretation which lead to the selection of the word place rather than space, or space rather than place, in everyday language and phrases. Put very simply, 'space' seems to indicate freedom, liberty, and opportunity, as indicated by common phrases such as "I need some space"; "when I get the space to work on *x* project"; "we're giving each other the space to think about what we want". In this sense, 'space', in contemporary common parlance, is viewed almost universally positively. 'Place', meanwhile, can be interpreted both positively ("this is the place for me"; "a place to fit in"; "I love this place") and negatively ("I know my place"; "this is no place for children"; "someone needs to put him in his place"). Thus, whilst space is almost always associated with freedom (Smith, 1994b), "place can also have repressive connotations" (Smith, 1994a, p253).

Although, as Mallett (2004) observes, it is important not to misunderstand 'place' and 'space' by reducing them down to a simplistic binary, several writers have been able to identify distinct characteristics and associations of 'space' and 'place' which demonstrate both the boundaries separating them (Morley, 2000) and the dialectical relationship between them (Waghorn, 2009; Withers, 2009). Thus, whilst space connotes the public, place connotes the private (Morley, 2000; Kidd and Evans, 2011); whilst space connotes movement, place connotes stasis and stability (Tuan, 1977; Waghorn, 2009; Withers, 2009); whilst space connotes the outside, place connotes the inside (Kearon and Leach, 2000); whilst space is objective, place is subjective (Cristoforetti et al., 2011). If each were capable of using language, space would use the spoken word whilst place would use the written word (de Certeau, 1984, cited in Waghorn, 2009).

When, then, does 'space' become 'place'? Withers (2009) argues that this happens as soon as a space has a name or a label attached to it. Others however contend that a space only becomes a place when it develops a specific layer of meaning, either through use and occupation (Waghorn, 2009), a series of social processes (Lefebvre, 1991), or emotional attachment (Tete, 2012). These dynamics which render space and place are clearly identifiable in the processes of creating home. Douglas (1991, p289) understands home to be "space under control", in that the rules of 'place' are applied to it.[21] Thus, the home is a space which is "enacted and occupied" (Waghorn, 2009, p277). This is applicable both in terms of home in the sense of built construction (Douglas, 1991) and homeland or territory (Tete, 2012), as both are a means of controlling and excluding people. Implicit in both is the role of feelings and emotion – as Smith (1994a, p272) puts it, home is the place "where sentiment and space converge". One of the key themes identifiable here though is the dialectic between public and private which is both mirrored and reversed in the dichotomy/dialectic between space and place. It is to conceptions of home as private space/place, and home as public space/place, and the harmony and tensions between them, that we will turn next.

Home as private space

Privacy is a phenomenon commonly associated with home in the literature (Oakley, 1974; Douglas, 1991; Oliver-Rotger, 2011), in spite of the fact that homes have not always been considered private space. Hareven (1991) contends that the notion of the private, exclusive home is essentially a bourgeois ideal[22] developed from the birth of capitalism. Prior to this period, homes were inherently social areas in the same manner as a pub or café (Morley, 2000), and the drift towards 'privatisation' of the home gathered pace alongside the Victorian preoccupation with eradicating social disorder. Thus, the private nature of the home came to be seen as the antidote to perceived social chaos, immorality, and lawlessness (ibid.). Therefore, the claim that privacy is implicit in home (Porteous and Smith, 2001) has not always been the case.

Contemporarily, the privacy of the home is deemed beneficial in a range of ways: as facilitating autonomy (Morley, 2000); as assuring sanctuary (Waghorn, 2009); as a human right (Fox, 2002; Waghorn, 2009); as encouraging creativity and intellectual thought (Oliver-Rotger, 2011), and as providing the conditions for individual identity to develop (Cristoforetti et al., 2011). However, conceptualising privacy in relation to home is not as straightforward as equating privacy with the home and publicity with the non-home – privacy can be further broken down and compartmentalised within the home setting (Douglas, 1991; Hobsbawm, 1991; Morley, 2000; Cristoforetti et al., 2011). Morley (ibid.) notes that the division of the home into separate rooms facilitates individual privacy – be it to fulfil bodily functions deemed to be intimate and, therefore, sequestered (Elias, 2000), or to facilitate individual pursuits – as Morley (2000) claims, "in an era of multi-set television households, personal computers, personal telephones and Walkmans, home can hardly be quite what it used to be".[23] Thus, on an individual level, "privacy is cherished in the home" (Douglas, 1991, p305).

The development of the view that bodily functions are private and should therefore be hidden is an interesting one. Elias (2000) provides an analytical account of the progress of social development, from *civilite* to civilisation, over a period of centuries. Drawing on a range of historical documents but focussing particularly on the work of renaissance scholar Erasmus,[24] Elias documents how notions of civility – what constitutes appropriate conduct in company or public forums – have changed over time. Thus, he is able to demonstrate that such bodily functions as urination, defecation, and the passing of wind, once accepted and even celebrated as expressions of bodily humanity have, through the civilising process, evolved into functions of a private nature, the hiding of which is constructed as modest and appropriate, civilised – whilst any public demonstration of such natural phenomena is condemned as shameful and immoral.

Elias (2000) notes that there are some negative consequences to the development of civilisation. Whilst civilisation has led to order and control, it has also resulted in individuals, some more than others, having restricted access to certain liberties. In short,

> socially necessary self-control is repeatedly purchased, at a heavy cost in personal satisfaction, by a major effort to overcome opposed libidinal energies, or the control of these energies, renunciation of their satisfaction is not achieved at all; and often enough no positive pleasure balance of any kind is finally possible, because the social commands and prohibitions are represented not only by other people but also by the stricken self, since one part of it forbids and punishes what the other desires.
>
> (Elias, 2000, p378)

The concept of *habitus* is crucial here. A term more commonly associated with medical descriptions of body type (Kagan, 1966),[25] Elias uses *habitus* to connote a way of being – a state of existence wholly created by one's social context. Here, the conduct, mindset, and way of life of an individual – *habitus* – is formed and controlled entirely by the nature of the individual's surrounding social environment – or, as Bourdieu (1985) would put it – field. What is significant, though, is the subtle, creeping, insidious manner in which *habitus* is developed, slowly, almost unknown, to the point where nobody really realises it is happening. As Bourdieu (1985) notes, *habitus* becomes fully entrenched when society has forgotten the genesis of the established norms; thus self-control – appropriate conduct as prescribed within the field – becomes "second nature" (Elias, 2000, p369). This is significant in that, prior to the development of Victorian ideals on public space,

> urban space lacked the forms of physical segregation which we now take for granted, in so far as distinct physical spaces were not necessarily reserved for particular activities – whether at the micro level, as between the different rooms in the house, or at the macro level, as between different districts of the city. For the reformers these various forms of heterogeneous mixing were seen as providing the threatening potential of pollution, destabilisation and disorder.
>
> (Morley, 2000, p21–22)

This does not mean though that the bathroom is the epitome of privacy within the home. Whilst the bathroom is for many considered an inviolable space when it is in use, Cristoforetti et al. (2011, p229) recognise the bedroom as the "inner sanctum" of privacy, whereby access is restricted to certain people at certain times. The bedroom, unlike the front doorstep, the sitting room, the parlour, or even the bathroom, epitomises privacy in that one only allows another into one's bedroom under very specific circumstances for very specific, albeit wide-ranging, reasons.[26] In this sense, the bedroom is the most private, and least public, area of the home, in that it is the most exclusive.

Relaxation is a good that privacy is thought to afford. As Smith (1994b, p32) puts it, "The home was seen as a place of regeneration, of repose and refreshment, and renewal of physical and psychological energy". However, the association between home, privacy, rest, and recuperation is not one that is equally applicable to all people (Stanko, 1985, 1990; Kidd and Evans, 2011; Tyner, 2012). Home affords rest and relaxation because of its domestic nature and the domestic processes that occur within it (Douglas, 1991). Reminiscent of school domestic science lessons, later renamed 'home economics', domesticity refers to cooking, cleaning, needlework, and other processes which make a household operate effectively (Murphy, 2011). Domesticity implies the ordinary, the everyday, the routine, the controlled, in sharp contrast with the strange, the unknown, and the untamed connotations of public space – to which we will now turn.

Home as public space

The concept of home is also applicable to public space, albeit in a different sense. The public space of one's home nation, hometown, or immediate neighbourhood, incorporates sensations of home that differ from the private inhabitation of the dwelling (Hill, 1991; Blunt, 2005; Kidd and Evans, 2011). The relationship between the individual and the public space where their home is located is one of identity, belonging, and a set of characteristics shared with others from the same location (Porteous and Smith, 2001; Mac an Ghaill and Haywood, 2011). Within this relationship is an implicit code of conduct which both dictates appropriate behaviour and facilitates the development of a locational culture:

> It is ... knowledge of the shared rules of interaction in public spaces that allows us to interact with strangers and acquaintances in relatively predictable ways. Breaches of these rules can feel upsetting or intimidating.
>
> (May, 2011, p116)

Here May is pointing to the notion of "civil inattention" – the acknowledgement of the presence of another in public space, without direct engagement – as the signifier of both a 'safe' public space and a civilised one.

Elias's (2000) analysis of the development of civilisation becomes more pertinent to the public/private space dichotomy in his conceptualisation of individual self-control of the body as an element of state control of the people via the monopoly of the legitimate use of force. Here, bodily functions are located within a spectrum of a range of physical urges including bodily drives towards acts of violence, and are controlled on two levels: "The controlling agency forming itself as part of the individual's personality structure corresponds to the controlling agency forming itself in society at large" (Elias, 2000, p373). Within what Elias (2000) terms "pacified social spaces", the monopolisation of force creates a social contract whereby as the threat of violence is diminished, so too is the opportunity to engage in violence, or other "possibilities of pleasurable emotional release" (ibid.). So, the state and social control that is exerted over expressions of violence extends to other forms of physical expression, like those associated with homeless and street-life people, "in the direction of a more continuous, stable and even regulation of drives and affects in all areas of conduct, in all sectors of life" (p374).

A key ideological shift having an effect here is the notion that more and more physical space should be privately owned rather than publicly accessible. There are deep-seated associations between public space and disorder: "Public space engenders fears, fears that derive from the sense of public space as uncontrolled space, as a space in which civilisation is exceptionally fragile" (Mitchell, 2003, p13). As he puts it:

> What makes a space *public* – a space in which the cry and demand for the right to the city can be seen and heard – is often not its preordained "publicness". Rather, it is when, to fulfil a pressing need, some group or another *takes* space and through its actions *makes* it public. The very act of representing one's group (or to some extent one's self) to a larger public creates a space for representation.
>
> (p37; emphasis in original)

So, for Mitchell, the power of public space and the platform of the public arena must be reclaimed and harnessed in order that the rights of all are upheld. When something is public rather than hidden, it is accepted, validated, included, and enfranchised, rendering it powerful.

Mobility versus stasis

To a large extent, the concepts of place and space are mirrored by the concepts of stasis and mobility. Stasis is the experience of being placed, or being held in place, whilst mobility is having the freedom, or space, to move at will (Smith, 1994b; Case, 1996; Morley, 2000; Mallett, 2004; Skeggs, 2004). Some commentators who conceptualise the private home[27] as female and the public home as male subsequently view stasis as an inherently feminine, passive state of being and mobility as synonymous with active, dynamic masculinity (Morley, 2000).[28] However, to understand place as stasis and feminine, and space as mobility and masculine is essentially reductionist, as the dialectic relationships between these concepts are complex and multi-faceted (Case, 1996; Mallett, 2004).

Both stasis and mobility can be understood as being valuable and desirable on one hand, and harmful and negative on the other. Turning first to stasis, or the state of being still and inactive (Morley, 2000), some clear positives can be attributed to the concept – stasis connotes a rooted and stable home (Fox, 2002) and a fixed point providing a sense of self-identity (Smith, 1994b; Porteous and Smith, 2001). Stasis can further connote, for the more privileged of the world, status (Fox, 2002), order (Porteous and Smith, 2001), respectability (Ronald, 2004), and social capital (Manturuk et al., 2010). Any sedentarism that arises from stasis can be offset by bringing the wider world to us via various media and technologies (Morley, 2000).

Yet stasis can also manifest itself in a negative manner, in the sense that it can indicate being 'stuck'; trapped through gender dynamics (Massey, 1994), race, ethnicity, and nationality (Sivanandan, 2001; Brun and Lund, 2012; Tete, 2012) or, relatedly, class position in the advanced capitalist global hierarchy (Morley, 2000; Skeggs, 2004). In this sense, "immobility increasingly acquires the connotation of defeat, of failure and of being left behind, of being fixed in place"[29] (ibid., p50). Stasis can occur in groups with more social capital through fear of 'mixing' with 'undesirables' (Morley, 2000) and "cripples and stifles" children and young people (Douglas, 1991, p288).

Mobility also implies different things in different circumstances. For some groups, mobility is an inherent good, as it indicates freedom away from the "tyrannies of the home" (ibid., p303), the benefits of travel (Porteous and Smith, 2001), and evidence of acceptance into (post) modernity (Tyner, 2012). As well as freedom of movement in a spatial sense, mobility also connotes *social* mobility through the accrual of social capital and augmentation of social relationships and social networks (Manturuk et al., 2010). Further, mobility can enhance the benefits and enjoyment of home via the creation of a dialectic between 'home' and 'away from home' which throws into relief the significance and value of home (Case, 1996). Again, though, mobility is a negative concept in other contexts; whilst nomadism is viewed by some as a 'good' afforded to some groups (Hill, 1991; Morley, 2000, Mac an Ghaill and Haywood, 2011), it is also a symbol of displacement, dislocation, and dispossession of peoples who are forced to leave their homes through war and conflict, 'natural' disasters, urban renewal, or global socio-economic processes (Cuba and Hummon, 1993; Smith, 1994a; Porteous and Smith, 2001; Bauman, 2004; Tete, 2012). In this regard, mobility frequently takes a downward trajectory, as dispossessed groups from refugees, to migrant workers, to "hobos"[30] (Datta, 2005, p537) are denied the status and ontological security that rootedness provides (Morley, 2000).

Thus, a complex climate has emerged whereby stasis and mobility are both good and bad, and the domain of both the enfranchised and the disenfranchised depending on the nature and the context of each, reflecting the dichotomies of both an active bourgeoisie and a 'stagnant' poor, and a stable bourgeoisie and a rootless, fragile, vulnerable poor (Morley, 2000). Further, in a world where mobility increasingly becomes a form of capital and, thus, the domain of the privileged (Skeggs, 2004), the inherent, 'natural' mobility of homeless or displaced people, due to them not having the designated appropriate site of sedentarism (i.e., a home), constitutes a challenge to the new rules of movement:

> The mobility of choice of the affluent British middle-classes, conducted in relative ease, is quite different from the mobility of the international refugee or the unemployed migrant … the continual emphasis on respectability, on becoming a 'good citizen' through self-protection, is also a move replicated in contemporary 'risk' literature, which suggests that the moral order

> has been spatialized in the production of particular classed selves figured in mobility debates.
>
> (Skeggs, 2004, p49)

In other words, as mobility and stasis have a dual meaning in terms of postmodern understandings of geography (Morley, 2000), it has become necessary to develop rules and restrictions on both mobility and stasis which allow groups enfranchised by status, wealth, and social capital the freedom to move and the freedom to stay put, whilst affording the disenfranchised no opportunity to do either. Mitchell (2003, p189), encapsulates this well in terms of restrictions on street homeless people:

> Anti-homeless laws have been challenged on the grounds that, by effectively banning some people from public space, they are in violation of homeless peoples' constitutional right to travel ... *our* mobility is predicated on the immobility of the homeless (emphasis in original).

The effects of loss of home, and the limited entitlement to (re) establishing home, on dislocated groups is considered in the next section.

Loss of home, homelessness, and domicide

For some academics, there can be no understanding of home without an understanding of homelessness, as one defines the other in binary opposition (Hobsbawm, 1991; Mallet, 2004; Datta, 2005; Kidd and Evans, 2011). Further, the home and homelessness dichotomy is reflective of other binaries – safe and unsafe, included and excluded, fearing and feared and, crucially in terms of an advanced capitalism climate, success and failure (Porteous and Smith, 2001; Datta, 2005; Kinsella, 2012). This section considers the effects of loss of home and the state of being homeless to provide a 360-degree conceptualisation of home and its varied meanings.

Just as the concept of home can be interpreted in a variety of ways, there is no one single meaning of homelessness, and some contention as to how it should be conceptualised. Edgar and Meert (2006) outline the European Observatory on Homelessness system of categorising homelessness, which groups people under four broad headings: rooflessness, i.e., sleeping in the open or in night shelters; houselessness, i.e., being in temporary or emergency accommodation; living in insecure housing, for example staying temporarily with family or friends, or with an abusive partner; and living in inadequate housing, which may mean unfit housing or in conditions of extreme overcrowding. However, policy-driven definitions such as this negate the level of nuance and complexity of the nature of homelessness, and how it is understood by people who have experience of being homeless.

It is perfectly possible to feel sensations of homelessness when one has a 'roof' to live under through, for example, lack of acceptance, unfamiliarity, or cruelty (Hill, 1991; Cuba and Hummon, 1993; Brun and Lund, 2012); equally, people who do not have a 'roof' deny that they are homeless, as other benefits commonly attributed to home are fulfilled (Tomas and Dittmar, 1995). Kidd and Evans (2011), in their research with street homeless young people, encountered various settings and processes which their respondents understood as home, including connections with networks of people, control over one's own life, and a content state of mind. Thus, a physical, permanent residence is not necessarily a pre-requisite for experiencing home, as indicated by one of their Native American respondents:

> To me, a home is where you lay your head. If [sic] has a roof over it … it doesn't matter if it doesn't have a door. It's where you lay your head. That's what you call home. And for my people, my Native people, home used to be a forest. That's what they called home. They didn't have no fucking houses or shelters, they just lived in what they had. That's basically what I do. Wherever I lay is my home. Any person that says any different is fucked. (p765)

Datta (2005, p555) uses the term "alternative domesticities" to encapsulate the ways in which home can be 'done' and 'performed', even in a 'homeless' setting. However, such renditions of home, whether they be in the street, the shelter, the travelling community, or the commune (Windsong, 2010), pose a challenge to contemporary understandings of home in that they are at odds with the notion of home as a consumerist commodity (Wright, 1991; Porteous and Smith, 2001; Fox, 2002). In part due to the prioritisation of home ownership[31] (Wright, 1991; Fox, 2002; Mallett, 2004), and in part due to growth in the home furnishing and decorating market dominated by retail giants like Ikea (Kling and Goteman, 2003; Aitken, 2004), home is now firmly entrenched in the advanced capitalist network of the production of wealth (in terms of housing as an investment) for home owners and the fluidity and growth of the home improvements market. This notion of home as property and a financial investment can detract from the authenticity of the home setting and, subsequently, diminish its meaning on an individual level (Wright, 1991; Smith, 1994a). Thus, home becomes confused and entangled with materialism and ownership of signifiers of success[32] (Datta, 2005); those who do not submit to this dominant discourse of home are constructed as an enemy to both capitalism and civility (Mitchell, 2003; Moore, 2008).

People 'lose' their homes as a result of a range of processes, including mortgage repossession (Fox, 2002; Neild and Hopkins, 2013), rising property rental rates, and stagnant wages (Standing, 2011; Reeve et al., 2016) marital breakdown (Randall and Brown, 2006), physical, sexual, and emotional violence (Hill, 1991; Tomas and Dittmar, 1995; Hagan and McCarthy, 1997; Newburn and Rock, 2005), addiction and substance misuse (Johnson et al., 1996; Hyde, 2005; Mallett et al., 2005), leaving the care system, prison, or the armed forces (Randall and Brown, 1994; Mendes and Moslehuddin, 2006; Cooper, 2013), natural and human-created disasters (Cuba and Hummon, 1993; Porteous and Smith, 2001; Brun and Lund, 2012), and urban renewal (Cuba and Hummon, 1993; Smith,

1994a; Porteous and Smith, 2001). Causes of homelessness tend to be 'push' factors (Mitchell, 2003; Grover, 2008) although some 'pull' factors do exist (Mac an Ghaill and Haywood, 2011).[33] However, these wide-ranging pathways into homelessness are not viewed as equal by the wider populace, and are variously responded to with pity, sympathy, empathy, charity, understanding, apathy, condescension, negativity, and criminalisation (Ravenhill, 2008). Arguably, though, it is the homeless people visible on the streets of city centres who suffer the greatest vilification (Anderson, 1999; Mitchell, 2003; Moore, 2008).

The variety of routes into homelessness are therefore distilled by those seeking to understand and conceptualise the 'problem' of homelessness, creating two opposing camps – the homeless campaigners or champions who seek to locate homelessness as a *structural* issue, and the apologists for structural inequality who seek to construct homelessness as solely the result of *individual* failings, as illustrated by Mitchell, (2003, p197):

> Instead of working toward a more just housing and shelter system … the official line is more geared toward demonising homeless people – making homeless people seem somehow less than human, endowed with fewer rights than those of us who live in houses. If there has been an overriding discourse about homeless people over the past decade, it has been that they are nuisances (or worse) to be rid of – pests and vermin who sap the economic and social vitality of the cities and the nation.

Thus, homeless people, particularly those visible in public areas, are constructed as an ideological 'enemy' which must be contained and neutralised. Moore (2008, p179) discusses how contemporary "street-life people" are characterised negatively for their propensity to "perform the whole range of social and physical activities in public places, including those which are generally regarded as private and/or inappropriate" (ibid., p180); anathema to the rules and customs of civilisation established long ago. Thus, street homeless people conduct all aspects of their lives (including those which propriety dictates should be kept indoors) in public space out of necessity rather than choice – they simply have no other alternative. Mitchell challenges such right-wing commentators as Robert Tier (cited in Mitchell, 2003, p16) who construct the public conduct of homeless people (sleeping, begging, and so on) as a 'lifestyle choice', the result of rational free will, pointing out that restricted access to the private domain coupled with limited resources in the form of, for example, public conveniences and temporary shelters, render any other recourse impossible.

Whilst Mitchell is right to stress that constructions of people electing to be homeless and to engage in street-life behaviour from a range of available options are both misleading and indicative of the ideological drive to deny homelessness as a social 'problem', the question of 'choice' in relation to homelessness raises some deeper issues. Clearly, many homeless people would choose not to be so if they had other options open to them, but so what if they did? Why does the idea of people choosing to appropriate public space in the manner that they do raise such concerns for the state, business, and the 'general public'? The contemporary drive towards the expansion of business is significant here. Drawing on classic Marxian thinking, Mitchell (2003) is able to demonstrate the contradictory position that homeless people are placed in with regard to advanced capitalist development – homeless people are an undesirable element in the *physical* sense, in that they are viewed as a phenomenon which prevents people from visiting (and, of course, spending money in) shopping districts; yet they are a necessary entity in the *theoretical* sense in that capitalist systems require a surplus population, or reserve army of labour, in order for them to work. Mitchell's (2003, p179) description of

> the stereotypical homeless person … a single white male skid row bum subsisting on mission charity and fortified wine. Considered misfits, wasted humans incapable because of their personal problems of realizing any part in the affluence of the postwar period guaranteed to all those who wanted it, they were perhaps to be pitied

is evocative of Merton's (1938, in Hayward and Morrison, 2005) retreatist, who struggles to cope with the conventional goals and means of capitalism and is thus incapable of enjoying the fruits of the boom. For Merton, the retreatist is both an inevitable and essential feature of a working capitalist system.

Thus, an inherent tension is presented, in that the existence of homeless people is essential, but there is no space for them to *be*. The answer, however, is found in legislative moves which continually restrict the freedoms of homeless people and their right to use public space. As Mitchell (2003, p174) puts it: "The homeless and poor are desperately *needed* but not at all *wanted*, and so the solution becomes a geographical one regulating space so that homeless people have no room to be *here*" (emphasis in original). Smith (1996) borrows the term 'revanchism' – developed from the French verb *revanchier*, to get revenge – from political science to provide a name for the political and commercial 'reclamation' of public space that has been thieved by groups perceived to pose a challenge to advanced capitalist constructions of security, morality, and family values in contemporary society. Mitchell (2003, p164) describes revanchism as "a right-wing movement of 'revenge' for the presumed 'excesses' of the liberal 1960s that seeks to revive what it sees as the 'traditional values' of America". With regard to homeless people in public space, revanchism has entailed the re-packaging and re-conceptualisation of homelessness as the result of individualised deviance, which has then acted as justification for punitive legal responses towards homeless people (ibid.).

Perhaps worse, though, than finding oneself homeless, is feeling that your home has been destroyed, purposely and deliberately, and can never be reclaimed. Drawing on his earlier conceptualisation of 'topocide' – the destruction or killing of place, Porteous, together with Smith (2001), has developed the concept of 'domicide' – the destruction or killing of home. The 'cide' suffix is crucial here, as it connotes murder; deliberate, pre-meditated, and wilful killing of home, in a calculated, cold-blooded manner with no possibility of mitigating factors (ibid.). Domicide can occur for a variety of reasons and in a variety of ways; however, Porteous and Smith (2001, p106) have identified two broad categories within which domicide can be understood – "everyday domicide" and "extreme domicide". 'Everyday domicide' is commonplace, frequent yet lengthy, and normalised; an example being the 'death' of Porteous' home village of Howdendyke in East Yorkshire:

> Its demise was begun by Thatcherite restructuring, the creation of unnecessary port installations to 'discipline' the unionised port of Goole nearby. Since 1969, port development has grown in several phases, each one of them taking more land once occupied by houses. Abandoned by planners and politicians as a 'dying village', Howdendyke is not dying, but rather being slowly killed off by piecemeal development. In 2001, after about 30 years of slow replacement, the village is only one-third of its original size, has lost much public access to the river, is heavily polluted, and has seen the demolition of its pub, shop and village hall.
>
> (ibid., p120)

'Extreme domicide' appears very different at first sight, as it results from, for example, natural or 'accidental' disasters, and affects the lives of individuals on a much bigger scale. Victims of extreme domicide are usually 'paid out' i.e., financially compensated for their loss very quickly, whereas victims of everyday domicide usually have to wait a long time for recompense or are denied victim status entirely (ibid.). However, Porteous and Smith note that environmental disasters often have a human cause, usually one that occurs through state/corporate activity, but is constructed as an 'act of God' to avoid the attribution of blame – a good example being the Aberfan disaster[34] (Kletz, 1993). Some such extreme disasters are used as a justification for implementing a new, usually punitive, economic system in the area, on the basis that the event has been so extreme, and so life-changing, that existing economic structures are no longer viable and must be started afresh. The affected populace, so shocked by events that they are rendered powerless in a state of suspended animation, are unable to resist the imposition of a new regime (Klein, 2007). The aftermath of Hurricane Katrina in New Orleans, characterised by the corporate assault on housing, healthcare, and education, is an excellent example[35] (Adams et al., 2009).

Although significant attachment to home is frequently "portrayed as sentimental and emotive, and therefore trivialised" (Fox, 2002, p586), domicide has a huge impact on those who experience it, tantamount to a bereavement, and responded to often with a long period of grieving (Brun and Lund, 2012; Tete, 2012). The process can impact on people's very sense of self and identity (Porteous and Smith, 2001). However, domicide continues to occur because it is constructed as being for the 'common good' or the 'greater good'; a 'shame' for the people who suffer it, but a necessary evil in the grand scheme of progress and development (ibid.). This presentation of domicide, particularly everyday domicide, as vital for the 'greater good', denies victim status for those who lose their home and renders them powerless in terms of preventing the domicide of their homes, as they are constructed as at best a nuisance, and at worst an 'enemy' standing in the way of a beneficial process. A pertinent example of this is the case of Capel Celyn, North Wales, a valley village which was purposefully flooded in 1965 to provide a reliable water supply for the growing population of Liverpool (Cunningham, 2007). The 50 residents of the centuries-old settlement, where Welsh was the first language, were displaced against their will with no possibility of ever returning as

> of all forms of domicide, the flooding of valleys behind dams is the most irreversible. Home is lost at several scales at once: dwelling, neighbourhood, village, land and landscape are obliterated ... most often, all physical props to memory are drowned beneath the surface of a permanent engineered lake.
> (Porteous and Smith, 2001, p136)

Thus, domicide can and does occur on a global scale and a local scale, affecting many people and few people, to a greater or lesser degree – incorporating everything from the appropriation of Diego Garcia as a US military base, because of its convenient location in the Indian Ocean, and the displacement of its population to nearby Mauritius (Vine, 2011), to the bullying of Glasgow residents from their homes to make way for a new stadium in the run up to the 2014 Commonwealth Games (Gray and Porter, 2014). Liverpool artist Jayne Lawless has documented, in various forms, the domicide of her family home, 150 Granton Road in the Everton district in Liverpool. Jayne's home fell victim to the Housing Market Renewal Initiative (HMRI); a New Labour-developed scheme designed to find 'solutions' to the problem of housing areas "on a downward trajectory towards area abandonment" (Webb, 2010, p313). HMRIs have been criticised on a number of levels – they over-promised on what they could achieve (Cole, 2012), they reflected a lack of clarity in terms of the 'urban sustainability' they sort to establish (Turcu, 2012) and, perhaps most significantly, they reduced the 'value' of home to a one-dimensional, monetary worth (Webb, 2010; Jones, 2013).

For Jayne Lawless the destruction of her home has been a drawn-out, deeply personal, and hurtful experience. In Jeanne van Heeswijk's short film, *Jayne Lawless in Granton Road*, Lawless details the deliberate ten-year process leading to the domicide: From the initial neglect and lack of maintenance, through promises of restoration with new walls, doors, and double glazing, to ultimately the compulsory purchase order[36] which forced her parents out of their home against their will and into a very similar house close by, for which they needed a new mortgage. The death of Lawless's mother shortly afterwards is, for Lawless, a direct consequence of the domicide. Lawless's story amply demonstrates that loss of home in this regard is much more than becoming homeless; it is a destructive process which has severe and long-lasting human impact, capable of damaging identity right to its very core.

Conclusion: Applying ideas about home to the Liverpool context

Clearly, then, home is a multi-faceted concept, which has been examined at length by academics, policy makers, and politicians and explained in different ways via a variety of different disciplines. Home as an idea, or as a feeling, extends far beyond the built construction we call a 'house', and is applicable as a concept on multiple levels, from the bed to the planet as a whole, via neighbourhood, city, and nation. As much as home is a noun, it is also a verb, as home is performed[37] on a day-to-day basis in all of the arenas in which it exists. Home is a concept that is both freeing and restraining and, crucially, home would not exist in the same way without the existence of homelessness and how it is constructed. Throughout this chapter, capitalist development is a recurring theme which impacts on home and how the concept has mutated to take different forms reflective of the advances of capitalism. The next chapter, Chapter Three, explores how home was first made in Liverpool, from the earliest stirrings of capitalism to the post-war period.

Notes

1 Although interestingly *aedes* in Greek means 'odious' or 'unpleasant', as in the aedes mosquito (Tyagi and Pande, 2008).
2 Ronald (2004, p57), however, notes that, in terms of understandings of 'house' and 'home' in Japan, "the concepts of family, house and household are bound together within the Japanese language as *'ie'*. In pre-war Japan the relationships defined in the organisation of *'ie'* constituted a legal organisational structure under which the paternal head of household held authority with a rule of one son succession.... In the post-war period, '*ie*' has been redefined in terms of expectations and obligations" (emphasis in original).
3 Kidd and Evans (2011) note, importantly, that the significance of having a place to sleep and rest in understandings of home extends beyond having a formal place of residence, in that a 'sense' of home can be achieved through having a place to 'lay your head' (p765).
4 Morley (2000, p17) points out that in some regards the construction of 'home' as an individual dwelling place is viewed as a negative attribute of social life, citing Benjamin Disraeli: "Home is a barbarous idea; the method of a rude age; home is isolation, therefore anti-social. What we want is *community*" (emphasis added).
5 Indeed it is the notion of neighbourhood and community that invokes more nostalgia and homesickness in migrants than any sense of nationhood (Mac an Ghaill and Haywood, 2011).
6 New Zealand's notion of the *marae*, or community home, is a similar concept (Waghorn, 2009).
7 Manturuk et al. (2010) note the significance of home ownership in evoking 'stakeholder' sensations in residents of a neighbourhood, and subsequent increases in perceived levels of social and cultural capital leading to perceptions of a 'good' neighbourhood. Similarly, Bone and O'Reilly (2010) identify the nature of the contemporary housing market in the UK as leading to the development of stark differences in the perceptions of 'good' neighbourhoods (characterised by home ownership) and 'bad' neighbourhoods' (characterised by rental properties and transience).
8 Morley (p52) gives a further example of this phenomenon from journalist Gary Younge: "for as long as I can remember, there was a tiny adhesive flag stuck to our front door.... It was the Barbadian flag and the rule was that whenever we entered the house we were not English – we were in Barbados and we would behave accordingly".

9 The sociological concept of meaning, closely associated with Mead's contribution to understandings of symbolic interactionism, locates the self in relation to the perceptions of others reflected on to the self (Fine, 1993).
10 Phenomenology is in essence a philosophical/sociological approach which understands experience to be essential to meaning (Cerbone, 2006).
11 "The transactionalist view…holds all behavioural experience involves the interaction of three things: people (individuals and groups), environment (physical and social), and time (linear and cyclical)" (Case, 1996, p2).
12 Skeggs (2004, p13) takes inscription to mean "the way value is transferred onto bodies and read off them, and the mechanisms by which it is retained, accumulated, lost or appropriated".
13 Waghorn's (2009, p275) useful discussion on New Zealand's "home invasion" legislation critiques the idea of violent crime carrying a greater penalty where the perpetrator has entered the victim's home, and cites a newspaper piece of the time when the Act was being debated which warned of the dangers of "equating real estate with flesh and blood" (ibid.).
14 The sensation of invasion is not provoked solely by uninvited intruders who have no legitimate mandate to be present in the home – Tamm (1991) notes that similar reactions are experienced by, for example, the elderly, disabled people, and those receiving care and rehabilitation after illness or injury towards care givers who have, to all intents and purposes, license to be present in the home milieu.
15 Lewicka (2005, p382), for example, notes that "place attachment may be a factor inhibiting mobility and individual progress".
16 Home ownership is thought also to enhance ontological security (Saunders, 1990); however, this viewpoint has been challenged; see e.g. , Fox (2002).
17 Smith (1994b) notes that women tend to draw a more direct link between security and home than men, despite the repeated finding that women are "more likely to be raped, assaulted and even killed at home than in any other place" (Goldsack, 1999, p123).
18 For example, "Home is for me yet a fortress from which to essay raid and foray, an embattled position behind whose walls one may return to lick new wounds and plan fresh journeys to further horizons" (Maxwell, 1963, cited in Porteous and Smith, 2001, p53).
19 Particularly the case for, for example, LGBT youth (Tyner, 2012).
20 Withers cites classic works by Lefebvre (1991) and Foucault (1980) amongst others as examples.
21 This is not to say, however, that there are no rules that are applicable to 'space', as we shall see elsewhere in the book.
22 As Hareven (1991) notes, the taking of lodgers grew to be frowned upon as a breach of middle-class family privacy.
23 Writing this sentence in 2000, Morley cannot have envisaged the explosion of individual home entertainment devices developed in the intervening years to further support his claim.
24 Erasmus of Rotterdam, as he came to be known, produced the far-reaching *De Civilitate Morum Puerilium* (On Civility in boys) in 1530 (Elias, 2000, p1).
25 In physiology, *habitus* is a commonly used term to connote body shape or build (Kagan, 1966).
26 For example, an adult may invite another adult into their bedroom to pay a visit to someone who is in bed due to ill health, or for sex, or simply for home maintenance purposes such as repairs or decoration. Whilst one might allow a passing acquaintance into the bathroom to use the toilet, or the kitchen for a drink of water, one would be unlikely to allow her/him into the bedroom without a clear motive. Children, on the other hand, are often compelled to open up their bedroom to friends and acquaintances due to limited control over and access to space within the home.

27 Many academics, including, for example, Elizabeth Stanko (1985, 1990), recognise that the private home is, for huge numbers of people, both a prison and a torture chamber, as the site of domestic and sexual violence. Tyner (2012) stresses that such violence is not gender-specific, and includes, for example, child abuse, elder abuse, and neglect.

28 Morley (2000) invokes notions of the male *flaneur*, traversing public space with freedom and impunity, in contrast to the notion of the home as 'feminine' space to the point that women are attributed a level of agoraphobia, such is their perceived entrenched connection to the private home.

29 Kidd and Evans (2011, p765) present the perspective of one of their homeless respondents: "I kind of have to make fun of people and their concept of home. It is shelter basically. Modern man [sic] traps himself [sic] in his [sic] home. You live in your shelter and there is no more outside world. It's just a scary place between work and home".

30 Datta (2005, p537) cites Anderson's (1999) classic study on 'hobos' – or migrant workers – and how their lives were characterised by outsider status and insecurity.

31 I refer here to ownership of property in the financial sense; the ability to feel ownership in the sense of a stake and an investment in a rented home is well established – see e.g., Hollows (2008).

32 Datta (2005, p553) discusses one of her respondents, Cashmere, and the way in which she conflates material possessions with both home and respectability: "For Cashmere, then, her route to 'home' was being carved by her domesticity in the shelter, by her accumulation of shoes and clothes, by 'doing the right thing' and by becoming a 'two-car family'.... By describing how her life had changed because of the car, she highlighted feelings of dignity and respect".

33 For example, Mac an Ghaill and Haywood (2011) note that rough sleeping in a perceived prosperous city, as a prelude to gaining employment, is preferable to having a roof in a place with limited prospects, as has been the case for many Irish men for decades, perhaps centuries.

34 The state crime referred to as the Aberfan disaster occurred in Wales in 1966 when a slag heap from a nearby colliery slid downhill and engulfed the local primary school, killing 116 children and 28 adults. Initially constructed as a tragic accident, there was a national outpouring of grief for the losses; however, Kletz (1993) highlights that the event was preventable, as landslides had occurred several times before; what had in fact happened was that the Coal Board had been negligent in their management of the site.

35 The Grenfell Tower disaster, caused by a faulty fridge-freezer and resulting in a fire which ravaged the building and killed 71 people, occurred on 14th June 2017. The actions of the local authority, Kensington and Chelsea, caused the fire to spread in the first instance, and appear to be encouraging a neoliberal resolution to the matter by constructing the area where the tower stands as prime real estate (Platt, 2017).

36 Compulsory purchase orders (CPOs) are a particularly unpleasant and drawn-out way to lose one's home, and often have a common pattern: An area will be identified as a 'problem' – perhaps because of the quality of the buildings, or perhaps because the buildings are 'in the way' – of a proposed road, sports stadium, shopping centre, or other project deemed to be for 'the greater good'. The owners of the properties in question will be offered an amount of money as compensation for their loss, but the amount is seldom commensurate with the original market value of the property, or enough to buy a similar property in a similar area. The longer one refuses to comply with the order, the more the financial offer decreases, as neighbours, fearful of losing out, succumb to the offer of compensation, and more and more buildings become derelict (Gray and Porter, 2014).

37 Boland (2010) discusses how home can be 'performed' in a spoken sense, in that a regional accent can be an indicator of a particular home neighbourhood/city/region. Sonic geography is the idea that sound, in this case a regional accent, can be understood

as a signifier of place and as an indicator of a shared identity. Boland identifies the 'Scouse' accent as key to a communal identity, an identity which extends far beyond Liverpool's geographical boundaries.

Bibliography

Adams V, Van Hattam T and English D (2009) 'Chronic disaster syndrome: Displacement, disaster capitalism and the eviction of the poor from New Orleans' *American Ethnologist* 36 (4) pp 615–636. doi:10.1111/j.1548-1425.2009.01199.x.

Aitken L (2004) 'Ikea: Chucking out the chintz around the world' *Campaign* 50 pp 17–18.

Anderson E (1999) *Code of the Street: Decency, Violence, and the Moral Life of the Inner City*. Norton: New York, NY.

Bauman Z (2004) *Wasted Lives: Modernity and Its Outcasts*. Polity Press: Cambridge.

Blunt A (2005) 'Cultural geography: Cultural geographies of home' *Progress in Human Geography* 29 (4) pp 505–515. doi:10.1191%2F0309132505ph564pr.

Boland P (2010) 'Sonic geography, place and race in the formation of local identity: Liverpool and Scousers' *Geografiska Annaler Series B Human Geography* 92 (1) pp 1–22. doi:10.1111/j.1468-0467.2010.00330.x.

Bone J and O'Reilly K (2010) 'No place called home: the causes and social consequences of the UK housing "bubble"' *British Journal of Sociology* 61 (2) pp 231–255. doi:10.1111/j.1468-4446.2010.01311.x.

Bourdieu P (1985) 'The genesis of the concepts of habitus and field' *Sociocriticism* 2 pp 11–24.

Brink S (1995) 'Home: the term and the concept from a lingusistic and settlement-historical viewpoint' in Benjamin D and Stea D (eds) *The Home: Words, Interpretations, Meaning and Environments*. Avebury: Aldershot, pp 17–25.

Brun C and Lund R (2012) 'Making a home during crisis: Post-tsunami recovery in a context of war, Sri Lanka' *Journal of Tropical Geography* 29 pp 274–287. doi:10.1111/j.1467-9493.2008.00334.x.

Campbell L and Parcel T (2010) 'Children's home environments in Great Britain and the United States' *Journal of Family Issues* 31 pp 559–584. doi:10.1177%2F0192513X09350441.

Carlen U and Jobring O (2005) 'The rationale of online learning communities' *International Journal of Web Based Communities* 1 (3) pp 272–295. doi:10.1504/IJWBC.2005.006927.

Case D (1996) 'Contributions of journeys away to the definition of home: An empirical study of a dialectical process' *Journal of Environmental Psychology* 16 pp 1–15.

Cerbone D (2006) *Understanding Phenomenology*. Acumen: Durham.

Charleston S (2009) 'The English football ground as a representation of home' *Journal of Environmental Psychology* 29 pp 144–150. doi:10.1016/j.jenvp.2008.06.002.

Cole I (2012) 'Housing market renewal and demolition in England in the 2000s: The governance of "wicked problems"' *International Journal of Housing Policy* 12 (3) pp 347–366. doi:10.1080/14616718.2012.709672.

Coolen H and Meesters J (2012) 'Editorial special issue: House, home and dwelling' *Journal of Housing and the Build Environment* 27 pp 1–10. doi:10.1007/s10901-011-9247-4.

Cooper V (2013) *No Fixed Abode: The Implications for Homeless People in the Criminal Justice System*. The Howard League for Penal Reform: London.

Cooper Marcus C (1995) *House as a Mirror of Self: Exploring the Deeper Meaning of Home*. Conari Press: Berkeley.

Cristoforetti A, Gennai F and Rodeschini G (2011) 'Home sweet home: The emotional construction of places' *Journal of Aging Studies* 25 pp 225–232. doi:10.1016/j.jaging.2011.03.006.

Cuba L and Hummon D (1993) 'A place to call home: Identification with dwelling, community and region' *The Sociological Quarterly* 34 (1) pp 111–131.

Cunningham M (2007) 'Public policy and normative language: Utility, community and nation in the debate over the construction of Tryweryn Reservoir' *Parliamentary Affairs* 60 (4) pp 625–636. doi:10.1093/pa/gsm037.

Datta A (2005) '"Homed" in Arizona: The architecture of emergency shelters' *Urban Geography* 26 (6) pp 536–557536–557. doi:10.2747/0272-3638.26.6.536.

De Beauvoir S (1997) *The Second Sex*. Vintage: London.

Despres C (1991) 'The meaning of home: Literature review and directions for future research and theoretical development' *Journal of Architectural and Planning Research* 8 pp 96–115.

Douglas M (1991) 'The idea of a home: a kind of space' *Social Research* 58 (1) pp 287–307.

Dovey K (1985) 'Home and homelessness' in Altman I and Werner C (eds) *Home Environments*. Plenum: New York, NY, pp 33–64.

Edgar B and Meert H (2006) *Fifth Review of Statistics on Homelessness*. FEANTSA: Brussels.

Elias N (2000) *The Civilising Process: Sociogenetic and Psychogenetic Investigations*. Oxford: Blackwell.

Felonneau M (2004) 'Love and loathing of the city: Urbanophilia and urbanophobia, topological identity and perceived incivilities' *Journal of Environmental Psychology* 20 pp 43–52. doi:10.1016%2FS0272-4944(03)00049-5.

Fine G (1993) 'The sad demise, mysterious disappearance, and glorious triumph of symbolic interactionism' *Annual Review of Sociology* 19 pp 61–87.

Fischer G (2001) 'External and shareable artefacts as opportunities for social creativity in communities of interest' in Gero J and Maher M (eds) *Proceedings of the Fifth International Conference 'Computational and Cognitive Models of Creative Design'*, University of Sydney, 9–13 December 2001, pp 67–89.

Foucault M (1980) *Power/Knowledge: Selected Interviews and Other Writings*. Panthenon: New York.

Fox L (2002) 'The meaning of home: A chimerical concept or a legal challenge?' *Journal of Law and Society* 29 (4) pp 580–610. doi:10.1111/1467-6478.00234.

Fox L (2005) 'The idea of home in law' *Home Cultures* 2 (1) pp 25–50. doi:10.2752/174063105778053445.

Galbraith M (1997) '"A Pole can die for the fatherland, but can't live for her": Democratisation and the Polish heroic ideal' *The Anthropology of East Europe Review* 15 (2) pp 119–139.

Giddens A (1990) *The Consequences of Modernity*. Polity: Cambridge.

Gillsjo C and Schwartz-Barcott D (2011) 'A concept analysis of home and its meaning in the lives of three older adults' *International Journal of Older People Nursing* 6 pp 4–12. doi:10.1111/j.1748-3743.2010.00207.x.

Goldsack L (1999) 'A haven in a heartless home? Women and domestic violence' in Chapman T and Hockey J (eds) *Ideal Homes? Social Change and Domestic Life*. Routledge: London, pp 121–131.

Gray N and Porter L (2014) 'By any means necessary: Urban regeneration and the "state of exception" in Glasgow's Commonwealth Games 2014' *Antipode* 47 (2) pp 380–400. doi:doi.org/10.1111/anti.12114.

Grover C (2008) *Crime and Inequality*. Willan: Cullompton.

Hagan J and McCarthy B (1997) *Mean Streets: Youth Crime and Homelessness*. Cambridge University Press: Cambridge.
Hareven T (1991) 'The home and the family in historical perspective' *Social Research* 58 (1) pp 253–285.
Hayward G (1977) 'Psychological concepts of "home"' *HUD Challenge* 8 (2) pp 10–13.
Hayward K and Morrison W (2005) 'Theoretical criminology: A starting point' in Hale C, Hayward K, Wahidin A and Wincup E (eds) *Criminology*. Oxford University Press: Oxford, pp 61–88.
Hill R (1991) 'Homeless women, special possessions, and the meaning of "home": An ethnographic case study' *Journal of Consumer Research* 18 pp 298–310.
Hobsbawm E (1991) 'Introduction' *Social Research* 58 (1) pp 65–68.
Hollander J (1991) 'It all depends' *Social Research* 58 (1) pp 31–49.
Hollows J (2008) *Domestic Cultures*. McGraw-Hill: Maidenhead.
Hyde J (2005) 'From home to street: Understanding young people's transition into homelessness' *Journal of Adolescence* 28 pp 171–183.
Jayne Lawless on Granton Road (2013) [video] Jeanne van Heeswijk. Available at https://vimeo.com/50310050. Accessed 12 June 2017.
Johnson T, Freels S, Parsons J and Vangeest J (1996) 'Substance abuse and homelessness: Social selection or social adaptation?' *Addiction* 92 (4) pp 437–445.
Jones A (2013) 'The silent protestor: Jayne Lawless' *The Double Negative*. Available at http://www.thedoublenegative.co.uk/2013/06/the-silent-protester-jayne-lawless/. Accessed 11 June 2017.
Jupp E (2013) '"I feel more at home here than in my own community": Approaching the emotional geographies of neighbourhood policy' *Critical Social Policy* 33 (3) pp 532–553. doi:10.1177%2F0261018313479011.
Kagan J (1966) 'Body build and conceptual impulsivity in children' *Journal of Personality* 34 (1) pp 118–128.
Kearon T and Leach R (2000) 'Invasion of the "Body Snatchers": Burglary reconsidered' *Theoretical Criminology* 4 (4) pp 451–472. doi:10.1177%2F1362480600004004003.
Kidd S and Evans D (2011) 'Home is where you draw strength and rest: The meanings of home for houseless young people' *Youth & Society* 43 pp 752–773. doi:10.1177%2F0044118X10374018.
Klein N (2007) *The Shock Doctrine: The Rise of Disaster Capitalism*. Penguin: London.
Kinsella C (2012) 'Re-locating Fear on the Streets: Homelessness, victimisation and fear of crime' *European Journal of Homelessness* 6 (2) 121–136.
Kletz T (1993) *Lessons from Disaster: How Organisations have No Memory and Accidents Recur*. Redwood Publishing: Melksham.
Kling K and Goteman I (2003) 'Ikea CEO Anders Dahlvig on international growth and Ikea's unique corporate culture and brand identity' *Academy of Management Executive* 17 pp 31–37. doi:10.5465/ame.2003.9474809.
Lalli M (1992) 'Urban-related identity: Theory, measurement and empirical findings' *Journal of Environmental Psychology* 12 pp 285–303.
Lewicka M (2005) 'Ways to make people active: The role of place attachment, cultural capital, and neighbourhood ties' *Journal of Environmental Psychology* 25 pp 381–395. doi:10.1016/j.jenvp.2005.10.004.
Lefebvre H (1991) *The Production of Space*. Blackwell: Oxford.
Lewicka M (2010) 'What makes a neighbourhood different from home and city? Effects of place scale on place attachment' *Journal of Environmental Psychology* 30 pp 35–51. doi:10.1016/j.jenvp.2010.10.001.

Mac an Ghaill M and Haywood C (2011) '"Nothing to write home about": Troubling concepts of home, racialization and self in theories of Irish male (e)migration' *Cultural Sociology* 5 (3) pp 385–402. doi:10.1177%2F1749975510378196.

Mallett S (2004) 'Understanding home: A critical review of the literature' *Sociological Review* 52 (1) pp 62–89. doi:10.1111%2Fj.1467-954X.2004.00442.x.

Mallett S, Rosenthal D and Keys D (2005) 'Young people, drug use and family conflict: Pathways into homelessness' *Journal of Adolescence* 28 pp 185–199. doi:10.1016/j.adolescence.2005.02.002.

Manturuk K, Lindblad M and Quercia R (2010) 'Friends and neighbours: Home ownership and social capital among low- to moderate-income families' *Journal of Urban Affairs* 32 (4) pp 471–488. doi:10.1111/j.1467-9906.2010.00494.x.

Massey D (1994) *Space, Place and Gender*. University of Minnesota Press: Minneapolis, MN.

May V (2011) 'Personal life in public spaces' in May V (ed) *Sociology of Personal Life*. Basingstoke: Palgrave, pp 109–120.

Meegan R and Mitchell A (2001) '"It's not community round here, it's neighbourhood": Neighbourhood change and cohesion in urban regeneration policies' *Urban Studies* 38 (12) pp 2167–2194. doi:10.1080%2F00420980120087117.

McCloskey R (2011) 'The 'mindless' relationship between nursing homes and emergency departments: What do Bourdieu and Freire have to offer?' *Nursing inquiry* 18 (2) 154–164.

Mendes P and Moslehuddin B (2006) 'From dependence to interdependence: Towards better outcomes for young people leaving state care' *Child Abuse Review* 15 (2) pp 110–126. doi:10.1002/car.932.

Mitchell D (2003) *The Right to the City: Social Justice and the Fight for Public Space*. Guildford Press: New York, NY.

Moore J (2000) 'Placing *home* in context' *Journal of Environmental Psychology* 20 pp 207–217. doi:10.1006/jevp.2000.0178.

Moore S (2008) 'Street life, neighbourhood policing and "the community"' in Squires P (ed) *ASBO Nation: The Criminalisation of Nuisance*. Policy Press: Bristol, pp 179–202.

Morley D (2000) *Home Territories: Media, Mobility and Identity*. Routledge: Abingdon.

Mulla S (2008) 'There is no place like home: The body as the scene of the crime in sexual assault intervention' *Home Cultures* 5 (3) pp 301–326. doi:10.2752/174063108X368337.

Murphy G (2011) 'The fall and rise of home economics education: newly available home economics archives at The Women's Library' *International Journal of Consumer Studies* 35 (5) pp 595–600. doi:10.1111/j.1470-6431.2011.01021.x.

Neild S and Hopkins N (2013) 'Human rights and mortgage repossession: Beyond property law using Article 8' *Legal Studies* 33 (3) pp 431–454. doi:10.1111/j.1748-121X.2012.00257.x.

Newburn T and Rock P (2005) *Living in Fear: Violence and Victimisation in the Lives of Single Homeless People*. Crisis: London.

Ng S, Kam P and Pong R (2005) 'People living in ageing buildings: Their quality of life and sense of belonging' *Journal of Environmental Psychology* 25 pp 347–360. doi:10.1016/j.jenvp.2005.08.005.

Oakley A (1974) *Housewife*. Penguin: London.

Oliver-Rotger M (2011) 'Homeplaces and spaces of their own' in Cisneros S (ed) *Critical Insights: The House on Mango Street*. University of California Press: Santa Barbara, CA, pp 285–325.

Platt E (2017) 'The looming tower' *New Statesman*, 6 October 2017, pp 30–34.

Porteous D and Smith S (2001) *Domicide: The Global Destruction of Home*. McGill: Montreal.
Randall G and Brown S (1994) *Falling Out: A Research Study on Homeless Ex-Service People*. Crisis: London.
Randall G and Brown S (2006) *Steps Off the Street: Solutions to Street Homelessness*. Crisis: London.
Ravenhill M (2008) *Culture of Homelessness*. Ashgate: Abingdon.
Reeve L, Cole I, Batty E, Foden M, Green S and Pattison B (2016) *Home: No Less Will Do: Homeless People's Access to the Private Rental Sector*. Crisis: London.
Relph E (1976) *Place and Placelessness*. Pion: London.
Ronald R (2004) 'Home ownership, ideology and diversity: Re-evaluating concepts of housing ideology in the case of Japan' *Housing, Theory and Society* 21 pp 49–64. doi:10.1080/14036090410033295.
Rykwert J (1991) 'House and home' *Social Research* 58 (1) pp 51–62.
Seamon D (1979) *A Geography of the Lifeworld: Movement, Rest and Encounter*. St Martin's Press: New York, NY.
Saunders P (1990) *A Nation of Home Owners*. Unwin Hyman: London.
Sivanandan A (2001) 'Poverty is the new black' *Race and Class* 43 (2) pp 1–5. doi:10.1177%2F0306396801432001.
Skeggs B (2004) *Class, Self, Culture*. Routledge: London.
Smith D (1994a) *Geography and Social Justice*. Blackwell: Oxford.
Smith N (1996) *The New Urban Frontier: Gentrification and the Revanchist City*. Routledge: New York, NY.
Smith S (1994b) 'The essential qualities of a home' *Journal of Environmental Psychology* 14 pp 31–46.
Somerville P (1992) 'Homelessness and the meaning of home: Rooflessness or rootlessness?' *International Journal of Urban and Regional Research* 16 pp 529–539.
Standing G (2011) *The Precariat: The New Dangerous Class*. Bloomsbury: London.
Stanko E (1985) *Intimate Intrusions: Women's Experience of Male Violence*. Routledge: London.
Stanko E (1990) *Everyday Violence: How Women and Men Experience Sexual and Physical Danger*. Pandora: London.
Tamm M (1991) 'What does a home mean and when does it cease to be a home? Home as a setting for rehabilitation and care' *Disability and Rehabilitation* 21 (2) pp 49–55.
Tete S (2012) '"Any place could be home": Embedding refugees' voices into displacement resolution and state refugee policy' *Geoforum* 43 pp 106–115. doi:10.1016/j.geoforum.2011.07.009.
Tomas A and Dittmar H (1995) 'The experience of homeless women: An exploration of housing histories and the meaning of home' *Housing Studies* 10 (4) pp 493–518.
Tuan Y (1977) *Space and Place: The Perspective of Experience*. University of Minnesota Press: Minneapolis, MN.
Turcu C (2012) 'Local experiences of urban sustainability: Researching housing market renewal interventions in three English neighbourhoods' *Progress in Planning* 78 pp 101–150. doi:10.1016/j.progress.2012.04.002.
Tyagi K and Pande B (2008) *Concepts of Biology*. Rastogi: New Delhi.
Tyner J (2012) *Space, Place and Violence: Violence and the Embodied Geographies of Race, Sex and Gender*. Routledge: New York, NY.
Vine D (2011) *Island of Shame: The Secret History of the US Military Base on Diego Garcia*. Princeton University Press: New Jersey.

Waghorn K (2009) 'Home invasion' *Home Cultures* 6 (3) pp 261–286. doi:10.2752/174063109x12462745321507.

Webb D (2010) 'Rethinking the role of markets in urban renewal: The housing market renewal initiative in England' *Housing, Theory and Society* 27 (4) pp 313–331. doi:10.1080/14036090903160026.

Weiman M and Wenju S (2007) 'Home: A feeling rooted in the heart' *Children's Literature in Education* 38 pp 173–185.

Windsong E (2010) 'There is no place like home: Complexities in exploring home and place attachment' *The Social Science Journal* 47 pp 205–214. doi:10.1016/j.soscij.2009.06.009.

Withers C (2009) 'Place and the "spatial turn" in geography and in history' *Journal of the History of Ideas* 70 (4) pp 637–658.

Wood D and Beck R (1994) *Home Rules*. Johns Hopkins University Press: Baltimore, MD.

Woodhouse M (2006) 'Making a new home: The importance of a home-like setting in nursing homes' *Sociological Viewpoints* 22 pp 103–110.

Wright G (1991) 'Prescribing the model home' *Social Research* 58 (1) pp 213–225.

3 Introducing Liverpool and the Liverpool home

In many ways Liverpool has always been a divided city, with the moneyed classes being segregated from the working classes, and the working class segregated both hierarchically, in terms of job, wages, and respectability status, and along lines of both religion and ethnicity (Pooley and Irish, 1984; Fagan, 2004; Pooley, 2006; Courtman, 2007; Taylor, 2009; Frost and North, 2013). This chapter will contextualise historically these competing conceptions of the Liverpool home. First, the housing and home making of the labouring classes in Liverpool will be explored, together with an assessment of the tensions within this transient and oppressed group. Second, the development of the other side of the Liverpool home – the home making of bourgeois merchant classes, will be considered alongside the notion of Liverpool as the "'second city' of the British Empire" (Wilks-Heeg, 2003, p43).

The entire *raison d'etre* for the city of Liverpool is its geographical location on the River Mersey, with direct access to the Irish Sea and, ultimately, the Atlantic Ocean (Lane, 1997; Aughton, 2008; Taylor, 2009; Rodwell, 2014). This prime and unique location allowed for the development of a system of wet and dry docks, facilitating the easy transportation of goods back and forth between Britain and the Americas and, consequently, the establishment of both a capitalist economic system and a 'British Empire' (Rodwell, 2014). Indeed, Taylor (2009, p13) notes that Liverpool is effectively a very early example of a "planned town … with [a] grid of six streets set out on a sandstone ridge alongside the tidal creek of the Mersey". In keeping with the requirements of a burgeoning capitalist system, the physical structure of Liverpool developed through the 17th, 18th, and 19th centuries in a binary format – with a dominant class of merchants, captains, bankers, and insurers housed in high Georgian grandeur, and a labouring class facilitating the smooth running of the docks and, by necessity, living in their shadows (Lane, 1997; Pooley, 2006). Thus, the Mersey served a crucial purpose in the rise of a capitalist, colonialist, exploitative Britain.

To facilitate the smooth running of the port, a cheap, continuous and, importantly, close-to-hand supply of dock labourers was a necessity (Pooley and Irish, 1984; Pooley, 2006); thus dock workers, many of them immigrants from Ireland and all of them employed on a casual (and therefore financially and socially precarious) basis, made their homes along the waterfront. This home making in the environs of the port was ostensibly advantageous to all concerned – dockers would hear by word of mouth that a ship had come in to port, and could make themselves available for work immediately (Pooley, 2006), thereby fulfilling the needs of the merchants in 'turning around' cargo quickly. Initially, housing in the dockland areas consisted of courts; dark, damp, back-to-back dwellings characterised by poor sanitation, overcrowding, and disease (Wilcox, 2011).

Perhaps because of the recognition that a steady supply of fit and healthy workers was a necessity,[1] 'slum' clearance and housing improvement in the dockland areas date back as early as the 1850s (Pooley, 2006); St Martin's Cottages, built in 1869 in the riverside district of Vauxhall, were the UK's earliest example of social housing (Wilson and Womersley, 1977; Taylor, 2011). However, Pooley (2006, p214) notes that, rather than these dwellings being reserved as "housing for the poorest poor" as the Housing Committee claimed, rental rates for St Martin's Cottages were equivalent to those for bigger, privately rented properties in the suburbs – thus rendering them out of the reach of most casual dock labourers and their families. Indeed, despite such early social housing initiatives, it was not until the 1930s that "for the first time the corporation was forced to house the genuinely poor" (Pooley and Irish, 1984, p68).

For Pooley (2006), this is indicative of a wider structural stance taken by the authorities whereby the housing picture in 19th-/early 20th-century Liverpool, whilst appearing from the outside to be progressive in its provision of social housing, was actually deliberately discriminatory in that the lowest paid and lowest skilled households were 'guided' into tenement blocks close to the port. These tenements, built specifically for the management of 'slum' clearance displacement (Pooley and Irish, 1984), were constructed to minimum specifications and, in the main, concentrated "close to the dockside to ensure a reliable labour force" (ibid., p109), a pattern that did not begin to change until the 1930s with developments in the public transport system.

Due to legislative change and the subsequent availability of funding for public provision of housing,[2] the Corporation of Liverpool embarked on a programme of house building – in the main two-, three-, and four-bedroom two-storey houses with gardens[3] – further towards the city outskirts (Wilson and Womersley, 1977; Pooley and Irish, 1984; Meegan, 1989; Pooley, 2006), with a concomitant stream of private sector house building developing alongside it (Couch, 2003). The explicit intention of the Corporation was that households would start out in the low-status tenements and, with the passage of time and increases in job prospects and/or wages, would move on to higher-status properties, either the public sector or the private sector, in the suburbs (Pooley, 2006).

However, as Pooley (2006, p216) observes, "the assumption was that through this process the older housing that was vacated would filter down to the poor, but in practice this rarely happened". Significantly, with regards to this failing in the assumed pattern of household progression, Pooley and Irish (1984, p90) are able to identify an implicit yet clear policy whereby the working classes who relied on social housing were stratified into a hierarchical structure, with the better-quality housing stock reserved, not only for the better paid, but also for those who managed to obtain higher social status:[4]

> In the eyes of the Housing Committee corporation tenants should not only be financially responsible, but also must adopt a lifestyle which was socially acceptable and which helped to protect the value of the corporation's property.

Thus, whilst it was assumed by the Housing Committee that tenants displaced due to 'slum' clearance would 'cope' better in a tenement flat than a house with a garden, those households moved into suburban houses were given a tenants' handbook of thinly veiled instructions to keep the place 'nice' (ibid.); thereby inscribing the working-class character with respectability for those who conformed, and exclusion for those who did not (Skeggs, 2004). In this way, the Corporation of Liverpool achieved a dual goal of a perception of duty of care to those most in need, and the further development of a capitalist system within the city:

> Whether housing policy actually improved the housing condition of those most in need was less important than the fact that those in authority were being seen to be doing something which both salved social consciences and fulfilled criteria of economic pragmatism. (Pooley and Irish, 1984, p109)

In this sense, both architects and town planners effectively assumed the mantle of the social engineer (Burnett, 1986).

Drawing a clear distinction between the respectable working classes and their disreputable counterparts (Lane, 1997) was possible in large part due to the demographic nature of the district where the majority of the casual dock labourers and their families were housed. Vauxhall, more specifically the Scotland Road area, was made up of a very transient population with links to rural Ireland (Pooley, 2006; Courtman, 2007; Aughton, 2008). The Irish potato famine had driven thousands of native Irish from their homes to Liverpool – many intending their stay in Liverpool to be a temporary stopgap before travelling on to New York, although a very significant number, due to a variety of reasons,[5] never made it beyond Scotland Road (Gallman, 2000).

In keeping with the colonial oppression of the indigenous population in Ireland by the British, and the negative commentary required to justify it (Dorney, 2014), the Irish population of the Scotland and Vauxhall wards was labelled as everything from drunken to backwards, criminal to seditious, lazy to revolutionary, dirty and diseased to feckless and immoral (Pooley, 2006). Effectively ghettoised, this population retained their own 'Liverpool Irish' dialect (Aughton, 2008) and were feared as a social evil, capable of contaminating the wider working-class population with their inherent inferiority (Pooley, 2006) and, importantly, their self-inflicted poverty (Courtman, 2007). The drive to build tenement dwellings close to the docks, and to segregate the poorest working-class households into such dwellings, can be read as a facet of the wider drive to contain the threat of the most unruly and least pliable sectors of the labouring classes to the development and growing prosperity of the city.

The tenements of the Scotland and Vauxhall wards and the version of home they offered were criticised in some quarters on many grounds: Lack of privacy; the travel of noise; the perception of a lack of play space for children; the potential for poor maintenance of communal areas; and, as the majority of tenement blocks had four or five storeys or 'landings', the stairs (Pooley and Irish, 1984). Large numbers of people living in very close proximity to each other facilitated the swift spread of infectious disease (Lane, 1997). Nonetheless, the tenements of Vauxhall and Scotland wards proved to be very popular and much beloved to many of those who lived in them, not least for the sense of community spirit they afforded to their inhabitants (Wilson and Womersley, 1977), and their very close proximity to the city centre (Fagan, 2004).[6]

Corporation dwellings established in the nascent suburbs of the city however, particularly those constructed during the 1920s under the 1919 legislation, were considered (at least in the abstract sense) as much more desirable properties (Wilson and Womersley, 1977; Pooley and Irish, 1984) and indeed preferable to the private sector developments in Everton,[7] Kirkdale, Walton, Anfield, Kensington, Wavertree, and Dingle, which gave rise to the view that "the townscape was ... dreary, consisting of 'a monotonous uniformity of terraced houses lining streets arranged in rectangular grids'" (Liverpool Interim Planning Policy Statement 1965; cited by Couch, 2003, p41). Yet the mass construction programme of low-density housing estates in the suburbs of Liverpool between 1919 and 1945 has resulted in "a high degree of internal variation" (Pooley and Irish, 1984, p11) due to a range of factors, including the fluctuating house building policies resultant from ever-changing housing legislation, the financial constraints of the time period, variations in housing allocation policy, and the changing fashions and trends amongst corporation tenants (ibid.). With the passage of time it has become apparent that the estates built in the 1920s under the more generous provisions of the 1919 Housing Act, for example, Walton Clubmoor, are perceived to be the best-quality corporation properties, and this is reflected in the pattern of applicants seeking tenancies there (ibid.).

Thus, a clear strategy can be identified whereby working-class corporation tenants were, from the outset, stratified hierarchically according to employment, economic, and social status; thereby rendering the pre-war working class effectively divided and segregated. However, the labouring classes in Liverpool were further divided along religious lines. Christianity manifested itself in two opposing forms – Protestant and Catholic – romanticised as 'the orange and the green' (Topping and Smith, 1977; Wilson and Womersley, 1977; Pooley and Irish, 1984; Pooley, 2006; Courtman, 2007; Frost and North, 2013). Again, rather than this division simply being a reflection of differing religious beliefs, rituals, and customs, it was inherently hierarchical and political. Protestant 'orange'[8] communities were aligned with Conservative party politics and principles, embodied in the political leadership of George Wise[9] (Muchnick, 1970; Meegan, 1990, Belchem and MacRaild, 2006); meanwhile the Irish Catholic communities of Vauxhall and Scotland wards aligned themselves with the emergent Labour Party[10] and Irish Republican politics.[11] The Irish 'underclass', caricatured as the "lowly Irish 'slummy'" (Belchem and MacRaild, 2006, p326), were essentially an oppressed immigrant group for whom Irish Republican politics offered a sense of status and collective identity[12] (Belchem, 2007). Indeed, the social, religious, and political hierarchy which prioritised 'orange' communities over 'green' was even reflected in the geography of the districts inhabited by each, with the overwhelmingly 'orange' area of Netherfield Road, cut into the sandstone of the high Everton ridge (Muchnick, 1970; Lane, 1997; Rogers, 2012), quite literally looking down on the portside dwellings of Scotland Road.

Liverpool was characterised by a further divide within the working class in that it has had for some centuries an established Black community[13] (Wilson and Womersley, 1977; Christian, 1997; Belchem and MacRaild, 2006; Murden, 2006; Courtman, 2007; Frost and North, 2013). Unlike post-Windrush era Black communities in other British industrial cities, Liverpool's Black population developed alongside the growth of the port and the African slave trade (Belchem and MacRaild, 2006). Christian (1997, p72) notes that the heritage of the Black community in Liverpool is both conflicted and muddied in that, from the outset, Liverpool's Black population was essentially a mixed race one, becoming established as Black sailors and freed slaves settled into family relationships with White women in Liverpool (Lane, 1997). Thus, Liverpool's Black populace, whilst facing the same processes of oppression and exclusion as those in other cities, endures another layer of oppression due to their singular circumstances, encapsulated in the (perceived by many to be negative) term Liverpool-born Black[14] (ibid.). As with the Irish immigrant community, oppression of Liverpool Blacks manifested itself spatially, with a ghetto effectively being established in the Granby and Toxteth areas of the city (Wilson and Womersley, 1977; Murden, 2006; Boland, 2008, 2010; Roberts, 2010). Thus, in this regard both the 'green' and the 'black' represented oppressed and segregated populations within the city[15] (Courtman, 2007).

Divided Liverpool

The story of Liverpool though has always been something of a tale of two cities, in that the lifestyle and home making of the workers who facilitated the smooth running of the port contrasts significantly from that of those profiting from their work – as Fagan (2004, p23) notes, this "clearly was poverty existing besides grandeur". The moneyed classes of Liverpool during the Georgian and Victorian eras displayed their wealth and status by building both "monumental edifices to trade and commerce" (Bainbridge, 1984, p85) in the centre of the city and elegant residential areas, initially in Abercromby (Wilson and Womersley, 1977) and subsequently further afield in Everton, Anfield, and Breckfield, which afforded sought-after views across the Mersey (Menuge, 2008), and the north of the city riverside area of Waterloo, where marine-style villas were fashionable (Burnett, 1986). The Abercromby area in particular was a popular and stylish place for merchants and their associates to set up home, and this district has achieved United Nations Educational, Scientific and Cultural Organisation (UNESCO) world heritage status contemporarily for its extensive intact Georgian buildings[16] (Heseltine and Leahy, 2011).

In a pattern which has echoes of the 'zones of transition' model developed by the Chicago School of Sociology,[17] it is apparent that the most wealthy in Liverpool developed a home making path which moved ever further from the centre of the city, with its chemical, moral, and human pollution, towards the outskirts of the city, higher up and with cleaner air, and along the waterfronts both north and south, where the benefits of living by the river were unspoilt by the side effects of industry, transport, and trade. Wilson and Womersley (1977) note that those who were able to afford it had already started to abandon Abercromby as early as 1920 in favour of marine villas better suited to the needs and status of the *haute bourgeoisie* (Burnett, 1986). Thus, the earliest examples of homes established by Liverpool port profiteers started out on the path to decay – a path which, hastened and encouraged by the bombing campaigns of the Second World War,[18] continued unabated for decades; the moneyed gentry establishing home territory further and further away from the docks, leaving properties behind which decayed to the point that only the poorest would inhabit them, bringing with them their associated social problems and negative reputations (Wilson and Womersley, 1977). Post-Second World War, the damage to Liverpool's built environment, coupled with obsolescence, or presumed obsolescence, of much of the city's housing stock, colluded with a variety of external forces to ensure that the Liverpool of the latter part of the 20th century would continue to change. This push towards change, presumed at the time to be change for the better, will be discussed in Chapter Four.

Notes

1 Pooley and Irish (1984) note that late 19th-/early 20th-century housing policy in the city reflects the tensions between the genuine concern of social reformers such as Elizabeth 'Bessie' Braddock and Eleanor Rathbone for the welfare of labouring classes, and capitalist drives to establish housing provision as a mechanism for wealth accumulation.

2 The 1919 Housing and Town Planning Act made provision for central government subsidies which facilitated better-quality construction, and reflected the standards recommended by the Tudor Walters Committee in terms of room size (Short, 1982).

3 Burnett (1986, p336) notes that the Housing Act 1919, which facilitated a widespread programme of corporation *house* (as opposed to tenement) building ensured that "the standard formerly achieved by the lower middle classes – the semi-detached house in a low-density development, the 'parlour', the kitchen instead of the scullery, the bathroom and the internal WC – was democratised to a much wider section of the working classes".

4 Indeed Pooley (2006) refers to comments made by 19th-century commentator and author Hugh Shimmin, expressing the view that people ultimately secure the quality of housing that they deserve, as indicative of the prevailing view of the time. Lane (1997, p101) writes of the same author, "Shimmin consistently wrote of the *other Liverpool* as if it were a separate society with its own social practices and codes of behaviour" (emphasis in original).

5 Some could not cover the passage to New York, and came to Liverpool with the intention of staying just long enough to earn the fare; many stayed because they never saved

enough to sail, or because they had developed attachments, or they quickly became ill with a contagious disease which left them unfit to travel (McRaild, 2011).

6 Fagan (2004, p6) discusses the benefits of living in a tenement block (Gerard Gardens) which effectively backed on to the Walker Art Gallery and the Museum of Liverpool (latterly renamed as the World Museum) – noting that "it was great having 'town' as your playground". Similarly, research undertaken by Balderstone et al. (2014) uncovered testimony from those brought up in south-central Liverpool who saw the city centre as a 'playground' during their childhood.

7 Lane (1997, p60), describes the Everton district as characterised by "geometrically neat rows of streets of new terraces conforming to regulations to promote public health".

8 Orange in this context refers to the Orange Order of Protestantism which reveres King William of Orange who defeated the Irish Jacobites at the Battle of the Boyne (Lenihan, 2003).

9 George Wise was a well-known and charismatic figure in the Liverpool Protestant Party, who had a significant following (Belchem, 2007).

10 Meegan (1990) notes that the Labour Party only really secured a political foothold in Liverpool from the 1950s onwards.

11 T. P. O'Connor, a member of the Irish Nationalist Party, was elected Member of Parliament for Scotland constituency and was in office between 1895 and 1929 (Belchem, 2007).

12 Belchem and MacRaild (2006, p326) note that "ethnic affiliation offered a sense of belonging that bridged socio-economic, gender and generational divisions, but strictly for those of creed and country". This sense of group identity, and the perceived need to protect it, very much manifested itself along 'us versus them' lines.

13 Rodwell (2014, p24) acknowledges that "Liverpool claims the oldest Chinese community in Europe and has long-established East African and Jewish communities" – however, neither have been subject to quite the same negative processes, nor the same level of academic enquiry, as the Liverpool-born Black community.

14 This term indicates a specific local heritage which does not incorporate the support and inclusion of an immigrant group with the same or similar ethnic background, and implies a mixed ethnicity dating back generations so much so that it is difficult to discern a clear ethnic heritage (Nassy Brown, 2005). Christian (1997, p70) suggests that "to try and pigeon-hole the essence of 'being Black' in a city such as Liverpool is complex and fraught with intellectual difficulty".

15 That is not to say that Black and Irish populations were inclusive towards each other; as is often the case, shared experience of oppression does not necessarily lead to solidarity – as Frost (cited in Belchem, 2006, p 64) asserts "the notion of 'Scouseness' was, and still is, something Black Liverpudlians are excluded from since to be 'Scouse' is to be white and working class".

16 The city currently has in excess of 1,500 listed buildings, many of which date back to the Georgian era (Aughton, 2008).

17 Proponents of this theory understand cities to be organic, living entities, whose inhabitants (particularly those with an immigrant background) move further and further away from the 'disorganised' city centre as they become more and more established (Hayward and Morrison, 2005, p71).

18 An excellent example being the original Custom House, destroyed during the 'May Blitz' of 1941 (Taylor, 2009).

Bibliography

Aughton P (2008) *Liverpool: A People's History*, 3rd ed. Carnegie Publishing: Lancaster.

Bainbridge B (1984) *English Journey, or, the Road to Milton Keynes*. Carroll and Graff: New York, NY.

Balderstone L, Milne G and Mulhearn R (2014) 'Memory and place on the Liverpool waterfront in the mid-twentieth century' *Urban History* 41 (3) pp 478–496.

Belchem J (2007) *Irish, Catholic and Scouse: The History of the Liverpool Irish 1800–1939*. Liverpool University Press: Liverpool.

Belchem J and MacRaild D (2006) 'Cosmopolitan Liverpool' in Belchem J (ed) *Liverpool 800: Culture, Character and History*. Liverpool University Press: Liverpool, pp 311–391.

Boland P (2008) 'The construction of images of people and place: Labelling Liverpool and stereotyping Scousers' *Cities* 25 pp 355–369. doi:10.1016/j.cities.2008.09.003.

Boland P (2010) '"Capital of culture – you must be having a laugh!" Challenging the official rhetoric of Liverpool as the 2008 European cultural capital' *Social and Cultural Geography* pp 627–645. doi:10.1080/14649365.2010.508562.

Burnett J (1986) *A Social History of Housing 1815–1985*, 2nd ed. Methuen: London.

Christian M (1997) 'Black identity in Liverpool: An appraisal' in Ackah W and Christian M (eds) *Black Organisation and Identity in Liverpool: A Local, National and Global Perspective*. Charles Wootton College Press: Liverpool, pp 62–79.

Couch C (2003) *City of Change and Challenge: Urban Planning and Regeneration in Liverpool*. Ashgate: Aldershot.

Courtman S (2007) 'Culture is ordinary: The legacy of the Scottie Road and Liverpool 8 writers' in Jones D and Murphy M (eds) *Writing Liverpool: Essays and Interviews*. Liverpool University Press: Liverpool, pp 194–209.

Dorney J (2014) *Peace after the Final Battle: The Story of the Irish Revolution 1912–1924*. New Island Press: Dublin.

Fagan G (2004) *Liverpool: In a City Living*. Countyvise: Birkenhead.

Frost D and North P (2013) *Militant Liverpool: A City on the Edge*. Liverpool University Press: Liverpool.

Gallman J (2000) *Receiving Erin's children: Philadelphia, Liverpool and the Irish famine migration 1845–1855*. University of North Carolina Press: London.

Hayward K and Morrison W (2005) 'Theoretical criminology: A starting point' in Hale C, Hayward K, Wahidin A and Wincup E (eds) *Criminology*. Oxford University Press: Oxford, pp 61–88.

Heseltine M and Leahy T (2011) *Rebalancing Britain: Policy or Slogan? Liverpool City Region – Building on Its Strengths: An Independent Report*. Department for Business, Education and Skills: London.

Lane T (1997) *Liverpool: City of the Sea*. Liverpool University Press: Liverpool.

Lenihan P (2003) *1690: Battle of the Boyne*. Tempus Publishing: Stroud.

MacRaild D (2011) *The Irish Diaspora in Britain 1750–1939*, 2nd ed. Palgrave: Basingstoke.

Meegan R (1989) 'Paradise postponed: The growth and decline of Merseyside's outer estates' in Cooke P (ed) *Localities: The Changing Face of Urban Britain*. Unwin Hyman: London, pp 198–234.

Meegan R (1990) 'Merseyside in crisis and conflict' in Harlow E, Pickavance C and Urry J (eds) *Place, Policy and Politics: Do Localities Matter?* Unwin Hyman: London, pp 87–107.

Menuge A (2008) *Ordinary Landscapes, Special Places: Anfield, Breckfield and the Growth of Liverpool's Suburbs*. English Heritage: Swindon.

Muchnick D (1970) *Urban Renewal in Liverpool*. The Social Administration Research Trust: Birkenhead.

Murden J (2006) 'City of change and challenge: Liverpool since 1945' in Belchem J (ed) *Liverpool 800: Culture, Character and History*. Liverpool University Press: Liverpool, pp 393–485.
Nassy Brown J (2005) *Dropping Anchor, Setting Sail: Geographies of Race in Black Liverpool*. Princeton University Press: Woodstock.
Pooley C (2006) 'Living in Liverpool: The modern city' in Belchem J (ed) *Liverpool 800: Culture, Character and History*. Liverpool University Press: Liverpool, pp 171–256.
Pooley C and Irish S (1984) *The Development of Corporation Housing in Liverpool 1869–1945*. Centre for North West Regional Studies: Lancaster.
Roberts L (2010) 'World in one city: Surrealist geography and time-space compression in Alex Cox's Liverpool' *Tourism and visual culture* 1 pp 200–215.
Rodwell D (2014) 'Negative impacts of world heritage branding: Liverpool – An unfolding tragedy?' *Primitive Tider* 14 pp 19–35. doi:10.1179/1756750515Z.00000000066.
Rogers K (2012) *Lost Tribe: The People's Memories*. Trinity Mirror Media: Croydon.
Short J (1982) *Housing in Britain: The Post-War Experience*. Methuen: London.
Skeggs B (2004) *Class, Self, Culture*. Routledge: London.
Taylor D (2009) *Liverpool: Regeneration of a City Centre*. BDP: Manchester.
Taylor K (2011) 'From the ground up: Radical Liverpool now' in Belchem J and Biggs B (eds) *Liverpool: City of Radicals*. Liverpool University Press: Liverpool, pp 159–171.
Topping P and Smith G (1977) *Government against Poverty? Liverpool Community Development Project 1970–75*. Social Evaluation Unit: Oxford.
Wilcox A (2011) *Living in Liverpool: A Collection of Sources for Family, Local and Social Historians*. Cambridge Scholars Publishing: Cambridge.
Wilks-Heeg S (2003) 'From world city to pariah city? Liverpool and the global economy' in Munck R (ed) *Reinventing the City: Liverpool in Comparative Perspective*. Liverpool University Press: Liverpool, pp 36–52. doi:10.5949/liverpool/9780853237976.003.0003.
Wilson H and Womersley J (1977) *Change or Decay: Final Report of the Liverpool Inner Area Study*. HMSO: London.

4 1950s and 1960s regeneration

Obliterating obsolescence and building a Liverpool for the future

The Nazi bombing campaigns of the Second World War had a significant and long-lasting impact on the built environment in Liverpool. The 'May blitz'[1] of 1941 in particular created extensive war damage which effectively left 'gaps' in configurations of buildings, perceived as war wounds or scars by the area's inhabitants (Rogers, 2010). Post-Second World War, a combination of factors ranging from a high post-war birth rate and shaky economic position (Muchnick, 1970) to local government lethargy and incompetence (Balderstone et al., 2014) colluded to ensure that many bomb sites would remain so for decades in some cases, to the point that debris became the everyday habitat of post-war children[2] and a source of frustration for those who grieved for the lost built environment and sought restoration (Rogers, 2010; Balderstone et al., 2014).

It was not until the late 1950s/early 1960s that any sort of post-war urban regeneration got off the ground in Liverpool. Couch (2003) notes that it was around this time that urban planning, or town planning as it was more commonly known, came to be seen as a profession and a craft in its own right, alongside the emerging conceptualisation of the city as a dynamic living organism. Thus, the town planning of the day was emblematic of a new dawn for urban life, very much reflecting the economic and social optimism of the 1960s (Balderstone et al., 2014), and focusing on a bright, clean future. Perhaps the best physical legacy of this futuristic, modernist optimism is St John's Beacon, built in the city centre in 1969 in an innovative space-age style (Muchnick, 1970) in sharp contrast to the surrounding 'dingy' Georgian and Victorian city scape (Tulloch, 2011).

The modernist tone of the Liverpool Interim Planning Policy Statement (1964) and the City Centre Plan (1965) harnessed this optimism and "civic confidence" (Murden, 2006, p402) to envision a new city centre which would acknowledge concerns of both practicality and physical beauty (Couch, 2003), and reject inner-city low-level living in favour of high-rise tower blocks[3] (Aughton, 2008; Rogers, 2010; Tulloch, 2011). The dominant theme of building skywards is for Tulloch (2011, p131) symbolic of the city's confidence:

> In the 1960s, additions continued to be made to the skyline as the city seemed to build ever upwards in a symbolic demonstration of masculine vigour,

> proclaiming to the world that Liverpool still had a lot to offer and those who thought the city was dead were in for a shock.

Thus, the future and prosperity of the city seemed assured, with new homes in the sky or in the suburbs fit for a new generation of workers employed in manufacturing at, for example, the newly established Ford Motor Company branch plant in Halewood – allowing for positive comparisons to be drawn between Liverpool and the original 'motor city', Detroit (Murden, 2006).

This period of positivity, prosperity, and hope in Liverpool coincided with one of the city's best-known eras, when "almost anything to do with Liverpool was 'news' in the 1960s" (Lane, 1997, p124), encapsulated in the term 'Merseybeat'. Described by Frost and North (2013, p9) as an "Indian summer", the early 1960s saw a series of 'pop groups' from the Liverpool music scene capture the zeitgeist of the moment and find international fame and cultural influence (Millington and Nelson, 1986; Aughton, 2008), which had the effect of bolstering the confident mood of a "new prosperous Liverpool" (Frost and North, 2013, p8). Arguably, the creative and commercial success of Liverpool during this time, coupled with the futuristic and optimistic building programme, both against a backdrop of global human progress and success epitomised by the excitement of the "space race" (White, 2015, p47), makes this period the zenith of forward-facing Liverpool, with great hopes for the future and uncertainty viewed as opportunity.

In order to facilitate the rebirth of the city that was anticipated through the lens of post-war optimism, it was of course deemed paramount that the city be cleansed of its antiquated, obsolete, war-damaged, built environment, particularly what was considered to be its 'unfit' housing stock (Couch, 2003). Conceptualised by Meegan (1990, p94) as "housing led regeneration", an extensive[4] programme of 'slum' clearance[5] was implemented, initially in the Everton, Toxteth, and Abercromby districts of the city (Muchnick, 1970), affecting one in seven of the population (Couch, 2003). Despite the developing recognition that cities are essentially dynamic entities which warrant both careful handling (Czarniawska, 2002) and scientific investigation (Couch, 2003), it is fair to say that very little 'planning' went into this "rapid large scale slum clearance" (Muchnick, 1970, p41).

Whilst Wilson and Womersley (1977) note that, given the size and speed of the clearance programme, very little attention was paid to management and general oversight, Muchnick (1970) argues that the whole clearance process was the victim of competing agendas and silo thinking within the local authority, prioritisation of the city centre above the districts, and, further, a political football. Specifically, Muchnick identifies a tension between town planners seeking to develop a "social renewal policy" (p13) which would take account of the notion of neighbourhoods as social networks, and a housing department under pressure to deliver dwellings fit for human habitation on a grand scale with limitations on both available land and funding. Thus, housing officers sought "to provide as many new houses as possible as quickly as possible" (p27) and viewed planners, with their concern for demography, community cohesion, and such 'niceties' as health facilities, schools, and play areas, as a barrier to housing progress. Added to this was the influence of local councillors who, in the knowledge that dispersal policies would impact on the local vote, campaigned out of self-interest to maintain their constituency, resulting in a

> multiplicity of separate interest groups and organisations, each having defined its own problems, each seeking different sets of goals, and each jealously guarding its own powers and prerogatives.
>
> (p20)

These 'slum' clearance and housing renewal processes effectively resulted in a long period of dereliction for parts of central Liverpool, both in terms of dilapidation of the built environment and neglect of affected communities. The limited budget available for the housing programme meant that costs had to be kept to a minimum, the net result being that building on land already cleared through bomb damage was prioritised (Muchnick, 1970) and, in the interests of efficiency, a great many habitable dwellings were bulldozed due to their proximity to 'unfit' dwellings[6] (Rogers, 2010). In this sense the 'value' of these districts, and the dwellings within them, was reduced to a monetary one (Couch, 2003). Further, the multiple issues hampering the building of new dwellings resulted in vast areas[7] being left as rubble in a state of "environmental anarchy" (Wilson and Womersley, 1977, p43), in some cases for many years (Rogers, 2010).

Perhaps, though, the dereliction of the communities fragmented and dispersed as a result of 'slum' clearance was even more extensive and severe (Muchnick, 1970; Wilson and Womersley, 1977; Murden, 2006; Rogers, 2010). Failing to take account of the "human side of renewal" (Muchnick, 1970, p67), the housing department paid little or no attention to the views of residents[8] or what outcomes were important to them, thereby disregarding the understandable desire to retain community spirit through preserving existing family, social, and religious ties (ibid.; Rogers, 2010). Indeed, planners and housing officers attributed a discourse of coarse, boorish ingratitude to dissenting residents:

> the people were seen as being the problem, just as much as their run-down houses were. We were ungrateful. We were uncivilised. We couldn't see beyond our home on the Hill.
>
> (Rogers, 2010, no page given)

At the same time, they failed to recognise what a difficult, distressing experience living amongst the debris for long periods could be, particularly for the elderly, who had

> strong ties with the neighbourhood, were not very mobile and needed small, ground floor accommodation nearby. … Five years elapsed between declaration of the compulsory purchase order and completion of rehousing. … As clearance and rehousing progressed the familiar surroundings became alien.

> Friends and neighbours left; landlords stopped doing repairs altogether. … The net result was the slow death of all established networks and values of the area.
>
> (Wilson and Womersley, 1977, p128)

Thus, many who witnessed the destruction of their home, whether in a speedy or a protracted manner, experienced a sensation of grief and a shift in the meaning of their lives, as points of reference were lost (Muchnick, 1970; Rogers, 2010). In this sense, just as the built environment that represented the home of these people was destroyed, so too were their social conventions – as a respondent of Rogers (2010, no page given) puts it –

> the breaking up of the Everton and Scotland Road communities was a disgrace and the people responsible should be brought up before a judge and charged with wrecking a way of life.

In essence, those displaced through 'slum' clearance generally found themselves with very limited choice in terms of where their new home would be and what form it would take, not least because of the housing department's policy of only allowing tenants to refuse an allocation twice before removing them from the waiting list[9] (Rogers, 2012). An identifiable pattern emerged whereby households would either be rehoused within the 'slum' clearance area they had come from in new 'high-rise' accommodation, or in a more traditional style two-storey property either in the outskirts of the city or in one of the region's new towns established during the 1960s (ibid.). Tower blocks like the monolithic Canterbury, Crosbie, and Haigh Heights in Everton were viewed as stylish, modern, and emblematic of a new dynamic era, "reflecting the architectural fashion of its time" (Wilson and Womersley, 1977, p51). Meanwhile, however, the existing multi-storey dwellings were increasingly viewed as obsolete:

> The 1930s tenements were hailed in their heyday as something of an architectural and social milestone in public housing. … But thirty years of neglect, poverty and overcrowding have taken their toll. The gardens which gave many of them their names have long since been replaced by tarmac.
>
> (ibid.)

Rogers (2010, no page given) notes that, in light of limited inner-city space, "there were only two solutions. One was to build high; the other was to build vast new estates on the outskirts". Several 'overspill' or "model estates" (Frost and North, 2013, p9) such as Kirkby,[10] Gillmoss, Croxteth, Fazakerley, Norris Green, Speke, Huyton, and Sparrow Hall were already in existence but were further developed during this period; others, like Cantril Farm,[11] Halewood, and Netherley were created to meet the post-'slum' clearance demand of the 1960s (Muchnick, 1970; Wilson and Womersley, 1977; Pooley and Irish, 1984; Lawless, 1989; Meegan, 1989, 1990; Murden, 2006; Pooley, 2006, Rogers, 2010, 2012; Frost and North,

2013). Constructed in a "ring of disparate, partially planned and spatially fragmented districts" (Murden, 2006, p413), and in keeping with the established housing department view that housing need only provide shelter (Muchnick, 1970), newly developed estates received their first inhabitants long before they could be considered 'ready' for them.[12] Indeed, as Muchnick (1970, p80) demonstrates, such infrastructural necessities as schools, transport, open space, and leisure facilities were at best secondary issues: "the Corporation's policy is in actual fact a housing programme coupled with hopes and prayers for ancillary development". Such disregard for the realities and necessities of living in these peripheral estates can, for Muchnick, simply result in the displacement of 'slum' conditions to other parts of the city rather than eradicate them.

Further afield than the peripheral overspill estates were the New Towns in the region – mark two (Skelmerdale, designated 1961; and Runcorn, designated 1964) and mark three (Warrington, designated 1968) (Wilson and Womersley, 1977), which, together with "expanding towns" such as Ellesmere Port, Widnes, Winsford, and Burnley (Evans, 1972, p184) took those displaced through 'slum' clearance a significant distance away from the homes they had left. Such developments, conceived of as post-war optimism, were viewed by both national and regional government bodies as harmonious with both the growth in car ownership (rendering distance less of a social concern) and changing demands regarding living space (Diamond, 1972). The "traffic-free layouts, open green spaces, trees, landscaping, new schools set in acres of playing fields and new airy buildings to work in" (Denington, 1972, p145) seemed particularly suited to the needs of young families cast in the role of "pioneers" (ibid., p146).

Roads, traffic, and transport in general were a key theme of both the Liverpool Interim Planning Policy Statement 1964 and the Liverpool City Centre Plan 1965 (Muchnick, 1970; Couch, 2003; Murden, 2006). For the city to be able to compete as a major player in the modern era, "the ambition was to reshape and redevelop what was perceived to be an obsolete and inefficient city centre" (Couch, 2003, p51). Thus, plans were drawn up to re-route existing streets, bring the M62 right into the city centre, build an elevated ring road around the immediate area, and pedestrianise much of the space within (ibid.). Ultimately, several of these ambitions were never fully realised – for example, the intricate, futuristic, warren-like system of elevated walkways, tunnels, and escalators for pedestrians in the city centre which was started but abandoned (Sharples, 2004). Other modernisation strategies, like the demolition of the dockside overhead railway (Balderstone et al., 2014), did go ahead. New road systems in particular were unpopular, both with local residents and the housing department, who feared that newly built properties would be sacrificed to make way for them[13] (Muchnick, 1970).

This prioritisation of roads over homes is perhaps best encapsulated in the case of the second Mersey tunnel, the Kingsway Tunnel, which opened in June 1971. To make way for the tunnel, vast swathes of the area surrounding Scotland Road, a main arterial route into the city centre from the north, were lost, leading to the displacement of thousands of residents (Meegan and Mitchell, 2001; Rogers, 2012). Compared somewhat euphemistically by Rogers to *Las Ramblas* in Barcelona, Scotland Road in the 1960s remained a densely populated area of the city, housing a mainly Roman Catholic, working-class population with close ties to the docks, and boasting a pub on virtually every corner.[14] Both Topping and Smith (1977) and Rogers (2012) recognise that, geographically and logistically, the entrance to the second tunnel would be best located to the south of the city centre; however, Vauxhall to the north was selected as the site, largely due to the area's relatively powerless position and structural inability "to mount effective protest"[15] (Topping and Smith, 1977, p27). The demolition and restructuring period was gruelling for many residents – not least the occupants of Lawrence Gardens, who found themselves marooned in the road system, surrounded on each side by a multi-lane carriageway (ibid.) – but it was the aftermath and the permanency of the tunnel entrance that proved to be the "body blow" (Rogers, 2012, p56) from which the area never recovered.

The impact of the destruction of the majority of the Scotland Road area arguably encapsulates the position of the ordinary people of Liverpool as the 1960s gave way to the 1970s – the home and way of life of residents was sacrificed in the quest for modernisation, and, significantly, these processes were constructed as in their best interests. Topping and Smith (1977, p25) contend that

> Vauxhall shows in acute and clear form, the negative effects of 'structural change'; change that is imposed from outside the area, the result of wider social and economic developments, and is in no way attributable to the inadequacies of local people or institutions, though the fact that it strikes here rather than somewhere else may reflect the area's weak bargaining position.

Thus, the ordinary Liverpool communities that entered the 1960s with optimistic hope for a brighter Liverpool life ended the decade markedly changed, through modernisation of their dwellings, through dispersal to peripheral estates and new towns, and through the prioritisation of modern modes of transport. Those sectors of the community who opposed or threatened this progress were pathologised and stigmatised as part of an emerging discourse of a problematic Liverpool which took firm root during the 1970s (Topping and Smith, 1977).

1970s: Recession, decline, regeneration gone wrong and the re-establishment of negative reputation

The early 1970s, whilst proving to be a difficult period for many British communities through economic recession (Dow, 1998), were positively catastrophic for Liverpool (Hayes, 1987; Meegan, 1990; Belchem, 2006; Roberts, 2010; Tulloch, 2011; Balderstone et al., 2014; Frost and North, 2013). The global recession of 1973 coincided with several other phenomena which wielded heavy blows on the city. Liverpool's geographical location on the Mersey, once so advantageous to the city's prosperity and growth, was now at the root of its downfall – the decline of the British Empire (Belchem, 2006), together with the UK's commitment to the European Common Market (Meegan, 1990; Murden, 2006; Tulloch,

2011) significantly reduced the amount of traffic and commerce through docks on the west of the country and, subsequently, the requirement for bulk processing industry[16] dockside (Taylor, 2009). Developments in the handling and transport of cargo, specifically the use of transport containers, rendered the docks to the south of the city obsolete[17] as only the Seaforth docks to the north of the city had the necessary physical characteristics to manage an increasingly 'containerised' system of cargo shipping (Belchem, 2006; Balderstone et al., 2014).

The demise of the docks and associated industries, which so many relied on for employment, were not however singular in this unfortunate series of events. Manufacturing industry in Liverpool has traditionally been characterised by what Heseltine and Leahy (2011, p22) refer to as "branch plant syndrome", whereby large corporations tended not to commit to the city and its workforce beyond a single plant rather than wholesale operations and headquarters – in part because of the perceived dominance of the service sector in the city (Robson, 1988; Couch, 2003), and in part because of an accepted view that the Liverpool workforce were both unreliable and "strike-happy"[18] (Murden, 2006, p431), leading to perceptions of the city as "England's most proletarian city" (Meegan, 1990, p93). Arguably these ideas about Liverpool and Liverpudlians meant that the city bore the brunt of the economic decline of the 1970s and its impact on business and industry, as firm after firm withdrew from their weakest point by closing down branch plants[19] (Meegan, 1990; Grady, 2014).

The net result of the decline of both the port and the already meagre manufacturing base in Liverpool was twofold: Mass unemployment, amongst a "population with little capital and few marketable skills" (Meegan, 1990, p91) and mass depopulation. Eversley and Begg (cited in Hayes, 1987) detail the causal reflexive relationship between the two; as a decrease in job opportunities leads to outmigration and a withdrawal of public/private investment, the negative effects of less workers and less consumers in a given area impact on retail, the service industry, and the housing market. Those in senior and professional roles move on, leaving behind posts which are difficult to staff, thereby deterring external investment and further embedding existing negative processes. Given that Liverpool had lost over 150,000 jobs by the late 1970s, affecting 20–30% of those of working age (Bichard, 2016), and the city's population had halved since the 1930s (Frost and North, 2013), it is fair to say that Eversley and Begg's analysis is particularly applicable here. Thus, by the 1970s the city had a post-industrial landscape characterised by decay, dereliction, and abandonment to match the urban debris left by the 1960s 'slum' clearance programme. Again, a discourse of self-inflicted demise was attributed to the population of Liverpool, deflecting attention away from the many and varied structural causes of decline, over which the city's inhabitants had no control (Topping and Smith, 1977; Belchem, 2006).

Aside from the seemingly irreparable damage caused to the city's working and economic life, the home life of the city continued to be disrupted and harmed throughout the 1970s. Many of the projects introduced so optimistically in the 1950s and 1960s ultimately proved to be both faulty and fragile in the more difficult social and economic climate of the 1970s (Meegan, 1990; Rogers, 2010,

2012; Balderstone et al., 2014). Life in new towns like Skelmersdale, and peripheral estates like Cantril Farm and Netherley did not, in reality, live up to the Corporation's sales pitch of "the new Jerusalem" (Davies, 2008) for a range of reasons. Promises made to keep existing communities together in a new location were often not kept (Broomfield, 1971), meaning that loneliness and isolation became significant problems, particularly for those housed in tower blocks due to low levels of footfall (ibid.). Rents in Cantril Farm were double those of 'slum' areas, and money had to be found to furnish newly attained properties (ibid.). Lack of amenities in the likes of Kirkby meant that children, limited in terms of free time options and accustomed to playing in rubble and debris, struggled to respect the built environment of their new homes (Rogers, 2010), leading to vandalism, graffiti, and destruction of property – particularly where cheap materials meant poorly constructed buildings (Davies, 2008). Netherley, by the time it was visited by Beryl Bainbridge in 1983, was so dilapidated that it looked to her as if

> it would make a good prison complex … . If the Russians came over here and made it into the subject of a documentary film, the Eastern Bloc would send food parcels and donations.
>
> (Bainbridge, 1984, p101)

As Meegan (2003, p56) synopsises, it became

> clear that [wholesale rehousing in peripheral estates and new towns] produced new problems of geographical and social dislocation for the 'decanted' families living in social housing perceived by them (particularly in its high-rise form) as soulless and 'anti-social'.

The experiences of residents who had been rehoused in their area of origin in many ways mirrored those of people moved outwards to the periphery and beyond. Much of the housing stock that was constructed during the 1960s and 1970s was of poor design and poor quality, and some was experimental in nature (Hayes, 1987). Some notorious examples of poor housing regeneration include the Radcliffe estate in Everton which, modelled on a 19th-century Cornish fishing village (Davis, 2002), was cramped, dark, and the cause of extensive fear of crime due to its covered walkways (Rogers, 2012); and Canterbury, Crosbie, and Haigh Heights, opened in Everton in 1965 and viewed as beacons of hope for the future, were by the 1970s universally known as 'The Piggeries' due to the extent of construction faults, vandalism, and neglect they experienced, becoming the centre of a well-known legal ruling,[20] used to explain a point of law to LL.B students (McKendrick, 2003).

Similarly, home life in existing, established areas of housing also suffered – plans to improve local shopping areas through funding for pedestrianisation and area enhancement in Walton, Anfield, and Tuebrook failed to come to fruition (Wilson and Womersley, 1977; Couch, 2003), areas like the Lodge Lane district were left in a state of limbo awaiting decisions about future developments

(Wilson and Womersley, 1977) and, despite emerging concerns for the preservation of the existing built environment[21] (Murden, 2006), modernisation in some cases spoiled rather than enhanced what was already there. As Bainbridge (1984, p93) put it, "Toxteth has been improved – that is to say vandalised – by the city planners and the architects". Thus, for many city inhabitants in all districts the "ruined Liverpool of the 1970s" (Balderstone et al., 2014, p484) led to a sense of abandonment.

Unsurprisingly, the ruin, dereliction, neglect, and despair apparent in the city's built environment was paralleled by angst, tension, and unrest within Liverpool communities (Wilson and Womersley, 1977; Couch, 2003). The difficult economic climate of the 1970s had brought extensive social problems to many British cities, effectively rendering them "disaster areas" (Hayes, 1987, p2); however, the specific issues facing Liverpool, and, importantly, their structural causes, seemed so singular in the Liverpool case as to render it unique (Wilson and Womersley, 1977; Lawless, 1989; Couch, 2003; Courtman, 2007; Allt, 2008). The 'different' nature of Liverpool's social problems manifested itself in several ways. Racial tensions had become commonplace in British cities as the children of 'Windrush era' grew to adulthood (Bowling, 1999); however, such tensions in Liverpool which spilled over into street fighting in the summer of 1972 (Wilson and Womersley, 1977), were further complicated via the mosaic-like and multi-layered identities of so-called Liverpool-born Blacks (Christian, 1997; Couch, 2003; Murden, 2006; Courtman, 2007). Black youth in particular had become increasingly ghettoised in the Liverpool 8 (Granby/Toxteth) district, and were subject to multiple levels of discrimination, ranging from poor employment prospects[22] to inscriptions of criminality, to the extent that "today white people are not encouraged to go into Liverpool 8 unaccompanied" (*The Daily Telegraph*, 15 January 1976; cited in Wilson and Womersley, 1977, p93).

The industrial decline of the city, in particular the demise of the docks, created a dispossessed and disenfranchised group of older homeless people. Having been a port and therefore to a large extent a transient city, a significant number of men were accustomed to inhabiting the city sporadically, reluctant or unable to develop established familial and social bonds (Foord et al., 1998). The demise of Liverpool's maritime project left "perhaps a thousand middle aged and elderly men in Liverpool drifting from furnished rooms to lodgings to hostels or dossing in derelict houses or doorways" (Wilson and Womersley, 1977, p31). At the other end of the age spectrum, school leavers, particularly boys who had expected to follow in the footsteps of their fathers and make a career, for example, on the docks, found themselves in the unexpected position of being unable to find work – especially those from the most socio-economically deprived districts of the city (Topping and Smith, 1977). A significant proportion of such young men, in a bid to kill time, turned to drugs; cannabis at first, then – through lack of knowledge of the difference between the two – heroin (Pearson and Gilman, 1994; Thornton, 2013), developing addictions which were fed in many cases through crime (Fazey, 1988; Meegan, 1989; Pearson, 1991). A further aspect of the multi-faceted 'unique' nature of 'problem Liverpool' arose as a result of Liverpool

Football Club's success in European football, which brought its supporters to European destinations where they obtained, either by purchasing or through looting, designer sportswear not available in the UK (Thornton, 2003). This led to a developing close association in the popular imagination between football, crime, violence, gang culture, and Liverpool (Lawrence and Pipini, 2016).

Thus, the specific manifestation of industrial decline in Liverpool from the 1970s was read to be *distinctive* – inherently different from the experiences of other British cities, resulting in different social problems and requiring different state responses. For Labour politician Margaret Simey,[23] the application of the 'different' label to the Liverpool 'problem' was indicative of the state's perspective that the city, rather than being a British city, was in effect more like a colony of the British Empire, complete with an aggressive indigenous population who refused to toe the line and, therefore, needed to be contained and neutralised (Bainbridge, 1984).

Arguably, one of the ways in which the state attempted to achieve this repression in the Liverpool case was via a series of 'social experiments' (Topping and Smith, 1977; Wilson and Womersley, 1977; Lawless, 1989; Couch, 2003; Meegan, 2003; Murden, 2006; Frost and North, 2013; Tallon, 2013). Lawless (1989, p5) notes that it was crucial that successive "governments were seen to be both caring and innovative" in the face of urban blight; thus, a series of experimental projects were "designed to explore specific problems of city life" (ibid.). Although these experiments were implemented in inner-city areas across the country (Topping and Smith, 1977; Tallon, 2013), it appears that Liverpool was examined more than any other as "since the 1960s, the city has been the recipient, or victim, of every urban experiment invented" (Parkinson, 1985, p16) and inner Liverpool in particular had "been surveyed remorselessly for … many years"[24] (Wilson and Womersley, 1977, p15). The first, the 'Social Malaise' study, threw up a wide range of social concerns including debt, access to libraries for children, marital fidelity, and personal hygiene (Topping and Smith, 1977); subsequent studies focused on more narrow issues e.g., Shelter Neighbourhood Action Project (housing) and Educational Priority Areas (schools) (Couch, 2003).

Much social experimentation of this sort is inherently positivistic in nature, in that it seeks to study a given problem from afar and attribute causal issues to the community where the problem manifests itself[25] and, arguably, this is what various government bodies sought to achieve (Wilson and Womersley, 1977). In the case of Liverpool, however, some studies began to take more of a 'standpoint' with regard to the populations they were surveying and the inherent difficulties they faced. The Vauxhall Community Development Project (CDP) 1970–1975 rejected the "social pathology approach to deprivation" (Topping and Smith, 1977, p6) and focused on neo-Marxist solutions (Tallon, 2013); meanwhile the Liverpool Inner Area Study (IAS) 1973–1976 recognised the role of prejudice and stereotypes:

> Codes evolve which, to the initiated, carry a wealth of meaning and prejudice … a bad reputation is one of the most difficult things to change as it will

> outlast the circumstances which gave rise to it. As such, it stamps the seal of disadvantage on inner city areas in a very powerful and intractable way.
> (Wilson and Womersley, 1977, p194)

Thus, the seeds of resistance to notions of pathology and negative reputations were sown to a large extent by researchers who rejected 'top-down' analyses and sought to empower communities with an empathetic approach (Lawless, 1989).

Whilst the list of experiments that the city was subjected to during the 1960s and 1970s reads like a litany of failure (Murden, 2006), some enjoyed success. The Scotland Road Free School, set up by two teachers disillusioned with the formal education system in 1970, prioritised child freedom and responsibility over order, discipline, and corporal punishment (Hake, 1975). Both founders were committed to social justice and the empowerment of working-class children, and their philosophy of non-hierarchical, child-led, democratic education was both valued by the children and respected by the wider Free School movement (ibid.). Another development that drew on the groundwork of the Vauxhall CDP and the Edge Hill/Granby/Kensington IAS was the emergence of authentic "working-class voices" depicting Liverpool life (Courtman, 2007, p194). In keeping with Bainbridge's (1984, p100) assertion that "Liverpool people have always been articulate and my family used words as though they were talking to save their lives", Bennett (2007) notes that the city has a long history of oral storytelling, ranging from naïve (i.e., outside of established 'appropriate' methods) poetry to political ballads (Courtman, 2007). Following the politicisation of communities through e.g., rent strikes (Topping and Smith, 1977), and in the wake of the 'social realism' turn of the 1950s (Millington and Nelson, 1986), groups like the Scotland Road Writers and the Liverpool 8 Writers formed to allow the oral history tradition to develop a more established, written structure and to provide a safe space for writers to share their work and explore their potential[26] (Courtman, 2007).

Pinto (2000, p89) discusses the work of Pierre Bourdieu in exploring the interface between knowledge and oppression, noting that, as an inherently politicised, campaigning academic, Bourdieu's premise is that "knowledge [should be] used for emancipatory aims". In a similar vein, Dittmer (2010) discusses the Foucauldian principle of 'no truth without meaning' in relation to knowledge, power, and attribution of 'culture'. Writers who emerged in Liverpool during the 1970s did so in the face of a range of obstacles, from constructions of working-class pride to restricted access to literary tradition and education (Bennett, 2007; Courtman, 2007). In conversation with John Bennett (2007, p230), playwright and author Willy Russell acknowledged the difficulties caused by a 'working-class' education for

> aspirant secondary modern kids, like me, who still talked with something of an accent but were nevertheless more concerned with things artistic/musical/ cultural than with the stereotypical football, fights and shagging.

Similarly, Courtman (2007, p205) discusses the impact of discourses of the working classes of Scotland Road as "stupid and thick" on poets and storytellers stymied by their ascribed class position. Therefore, members of the Scotland Road and Liverpool 8 writers groups told their stories both in the face of adversity and as a form of "creative resilience" (ibid., p208) and political resistance.

Charlesworth (2000, p51) draws on Merleau-Ponty's concept of "primacy of perception" to suggest that those who live in poverty cannot detach themselves from their impoverished, deprived position; that poverty becomes the filter through which every aspect of life is viewed:

> There is ... a deep primacy of perception in the day-to-day experiences of working class people: a shaping, undeniable force that destroys for them any possibility of being other than they are and which sets parameters to their ways of dealing with the world.
>
> (ibid., p52)

Courtman (2007, p206) however identifies the inherent tension in writing from the oppressed working-class perspective, citing a respondent who pointed out that

> some of the early books [published by the Federation of Working Class Writers and Community Publishers] created a lot of controversy ... family members would come along and say, 'You shouldn't have written that. You shouldn't pull out skeletons in the cupboard'. You know ... when people describe very graphic poverty, other family members come along saying, 'You don't parade your poverty – hide it'.

Thus, it can be argued that the 1970s proved to be simultaneously a period of destruction and creation in the Liverpool working-class setting – the existing ways of living were harmed and diminished through a range of ruinous processes, yet in other ways working-class voices, passion, spirit and, crucially, *culture* were gaining more confidence, more exposure, and more recognition. Whilst many in the city remained within a shell of ignorance, preferring to "deaden the self" (Charlesworth, 2000, p59) as full and honest acknowledgement of their predicament would prove too painful, others engaged in a Freirean[27] scholarship which would both elucidate and emancipate whilst at the same time exposing the notion of Liverpool life and Liverpool culture to wider audiences (Millington and Nelson, 1986; Bennett, 2007; Courtman, 2007; Allt, 2008; Boland, 2008). Chapter Five continues to explore these processes, and the impact they had in changing the trajectory of Liverpool's future.

Notes

1 The period between 8–15 May 1941, known as the 'May blitz', saw the city subjected to "a blitzkrieg: 2,315 bombs, 119 landmines and countless incendiaries [being] dropped" (Taylor, 2009, p16).

2 Balderstone et al. (2014, p485) discuss how their research exposed the impact of war damage on children's play: "Younger interviewees remembered the city centre as a playground, full of open spaces, rubble, ruins and monumental dock infrastructure to climb on. The site of the Custom House, damaged in the war and then demolished in the late 1940s, was not filled in for years afterward .... Children imposed their own meanings on the ruins to suit their games, interpreting the barred windows as cells that once housed smugglers, whereas in reality they were designed to keep people out rather than in".

3 Tower blocks also offering the benefit of being cheap and quick to construct, and having a much smaller 'footprint' on potentially scarce and expensive land (Rogers, 2010). The physical and social problems associated with high rise urban accommodation are well documented – see e.g., Church and Gale (2000) for an overview.

4 Hayes (1987) notes that 33,000 'unfit' dwellings were demolished between 1966 and 1974.

5 I have read extensively on the Liverpool 'slum' clearance programme of the 1950s/1960s and have never seen any attempt made to qualify what is meant by the term 'slum' or challenge it in any way – it appears to be wholly accepted in a positivistic manner. The closest thing I have found to a critical interrogation of the term is Muchnick's (1970, p28) recognition that "legally, slum is a physical concept" i.e., factors beyond its built construction are not considered within law.

6 Ken Rogers (2010, no page given), who grew up in Everton before being rehoused in Norris Green, questions the need to demolish so much of the Everton area when "the adjacent community of Anfield remained virtually untouched, and the vast majority of those little houses are still going strong, fifty years on".

7 In the case of Everton, significant stretches of land cleared during the 1960s never were used for new housing – they were left abandoned for many years before being 'landscaped' and re-named Everton Park during the 1980s (Rogers, 2010).

8 The opinions and desires of residents were well known to planners, politicians, and housing officers, in part due to the work of Area Community Wardens whose 'educational' role within communities facilitated ground level contact with, and understanding of, the needs of the community (Muchnick, 1970). Indeed, Liverpool's Housing Director in the 1950s, Dr Ronald Bradbury (cited in Rogers, 2010, no page given), stated "they do not want to go to Kirkby or Speke. What they want is an up-to-date, super-duper dwelling just across the street".

9 The "'three strikes and you're out' rule terrified many people and forced rash decisions" (Rogers, 2012, p179).

10 Pickett and Boulton (1974) note that different estates were built for different reasons: Kirkby was developed predominantly to solve the problem of overcrowding in inner Liverpool, whilst Halewood, five years later, predominantly housed those displaced via 'slum' clearance. The big families rehoused in spacious dwellings in Kirkby led to the area earning the nicknames of 'Bunnytown' and 'Kidsville' (Meegan, 1989), and a further estate, Tower Hill, being completed in 1974 as an "overspill of the overspill" (Meegan, 1990, p95), for the children of the first Kirkby 'settlers' in Southdene.

11 Although Meegan (1989) notes that the farmland on which Cantril Farm would later be built was actually purchased by Liverpool Corporation before the outbreak of the Second World War for the purpose of overspill house building, the war being the reason for the delay.

12 Cantril farm, for example, did not have a single public phone box (Muchnick, 1970); Sparrow Hall, despite housing a predominantly Roman Catholic population, had no Roman Catholic school close by (Pooley and Irish, 1984) and Fazakerley felt so underdeveloped to one former Everton resident that "it was as if we had moved to the country" (Rogers, 2012, p190).

13 Meegan and Mitchell (2001) and Rogers (2012) note that buildings that had been standing for just four or five years were cleared to make way for newly created roads.

14 By the late 1960s, the Scotland Road area was well known to be suffering from entrenched socio-economic deprivation; further, residents had a keen awareness of their structural position. Topping and Smith (1977, p95) discuss how the *Scottie Press*, a local newsletter developed by Vauxhall Community Development Project (see page 95) "contrasted the money spent on the tunnel with the squalor in the area. An attempt was made to get the Queen, who opened the tunnel, to drive though the area. This proved unsuccessful".

15 There was in fact significant local resistance to the location of the tunnel entrance, including a Homes Not Roads campaign, which was ultimately unsuccessful (Rogers, 2012).

16 Tate and Lyle's sugar refinery struggled on until closure in 1981, with British American Tobacco pulling out soon after (Tulloch, 2011).

17 Balderstone et al. (2014) note that the majority of 'south end' docks were closed by 1972, with the last, Brunswick Dock, closing in 1975.

18 Murden (2006, p410) details how the perceived "independent culture and mindset of the docker" meant that "firms such as Ford actively discriminated against the very young, dockers, seamen, the unemployed or anyone from industries with a history of trade union activity". Lane (1997, p72) notes that ex seamen in particular were seen to embody "a distinctive set of habits and an outlook that harmonised with a community accustomed to uncertainty and casual labour", meaning that "moments were to be seized and wrung dry" (ibid.).

19 Bichard (2016) notes that over 350 factories closed in Liverpool between 1966 and 1977, including such famous names as English Electric, Dunlop, Huntley and Palmer, Crawfords, Fisher-Bendix, Meccano and Bibby's, (Friend and Mitchell, 1981; Sinclair, 2014).

20 In the case of *Liverpool City Council* v. *Irwin 1976/1977*, Liverpool City Council took Mr and Mrs Irwin, residents of Haigh Heights, to court for non-payment of rent. The Irwins brought a counter claim against the authority for failure to maintain common areas, for example, the stairwells and lifts, which were in a very poor condition. It subsequently came to light that the tenancy agreement mentioned only the responsibilities of the tenants; no mention was made of the responsibilities of the landlord. The case ultimately went to the House of Lords and it was established that Liverpool City Council had an obligation to maintain the common areas implicit in the agreement (McKendrick, 2003).

21 Between 1974 and 1978 a total of eight 'conservation areas' were identified: Castle Street (Pier Head), William Brown Street, Rodney Street, Canning Street, Princes Road/Princes Park, Sefton Park, St Michael's Hamlet, and Fulwood Park (Couch, 2003). It is worth noting that all of these areas are either in the city centre or further south – the north end of the city, home to Walton Park, Stanley Park, and both of the city's main football stadia, was evidently not considered worth conserving.

22 It quickly became apparent to researchers working on the Liverpool Inner Area Study that young Black people in Liverpool were subject to extreme prejudice when seeking employment, so much so that part way through the study researchers published a consultation paper, titled 'No one from Liverpool 8 need apply' – a statement taken verbatim from a job advertisement in a shop window (Wilson and Womersley, 1977, p22).

23 Margaret Simey was elected to the local council to represent Granby ward in 1963, became chair of Merseyside Police Authority in 1981 and, throughout her life, was a champion of the Toxteth area and its people (Clarke, 2004).

24 Such experiments in the Liverpool area ranged from national projects like Educational Priority Areas, Community Development Projects, and Inner Area Studies, to specific local projects like Urban Programmes, Inner Area Partnerships, Shelter Neighbourhood Action Projects, and Brunswick Neighbourhood Project (Topping and Smith, 1977; Couch, 2003; Murden, 2006).

25 See e.g., Topping and Smith's (1977) discussion on the difficulties of engaging academics from the universities of both Liverpool and Oxford in the structural causes of community strain in the Vauxhall area.
26 Celebrated Liverpool writer Willy Russell (cited in Bennett, 2007, p237) reflects on his working-class educational background and notes that he "was someone who felt that language had (for complex cultural reasons) been denied me. Maybe that's what created within me my hunger for language. For me language was power".
27 In *Pedagogy of the Oppressed* (1972), Paolo Freire argues, from a Marxian perspective, that class consciousness and subsequent liberation starts with the processes of recognising, documenting, and disseminating the experience of oppression.

Bibliography

Allt N (2008) 'The un-English disease' in Allt N (ed) *The Culture of Capital*. Liverpool University Press: Liverpool, pp 13–42.

Aughton P (2008) *Liverpool: A People's History*, 3rd ed. Carnegie Publishing: Lancaster.

Bainbridge B (1984) *English Journey, or, the Road to Milton Keynes*. Carroll and Graff: New York, NY.

Balderstone L, Milne G and Mulhearn R (2014) 'Memory and place on the Liverpool waterfront in the mid-twentieth century' *Urban History* 41 (3) pp 478–496. doi:10.1017/s0963926813000734.

Belchem J (2006) *Merseypride: Essays in Liverpool Exceptionalism*. Liverpool University Press: Liverpool.

Bennett J (2007) 'Manners, mores and musicality: An interview with Willy Russell' in Jones D and Murphy M (eds) *Writing Liverpool: Essays and Interviews*. Liverpool University Press: Liverpool, pp 228–238. doi:10.5949/upo9781846314476.017.

Bichard E (2016) 'Liverpool case study' in Carter D (ed) *Remaking post-industrial cities: Lessons from North America and Europe*. Routledge: Abingdon, pp 153–170.

Boland P (2008) 'The construction of images of people and place: Labelling Liverpool and stereotyping Scousers' *Cities* 25 pp 355–369. doi:10.1016/j.cities.2008.09.003.

Bowling B (1999) *Violent racism: Victimisation, policing and social context*. Clarendon: Oxford.

Broomfield N (1971) *Who Cares?* Documentary. Available at http://www.liverpoolpictorial.co.uk/blog/liverpool-1971/. Accessed 30 March 2016.

Charlesworth S (2000) 'Bourdieu, social suffering and working class life' in Fowler B (ed) *Reading Bourdieu on Society and Culture*. Blackwell: Oxford, pp 49–64. doi:10.1111/j.1467-954x.2001.tb03532.x.

Christian M (1997) 'Black identity in Liverpool: An appraisal' in Ackah W and Christian M (eds) *Black Organisation and Identity in Liverpool: A Local, National and Global Perspective*. Charles Wootton College Press: Liverpool, pp 62–79.

Church C and Gale T (2000) *Streets in the Sky: Towards Improving the Quality of Life in Tower Blocks in the UK*. Birmingham National Sustainable Tower Blocks Initiative: Birmhingham.

Clarke R (2004) 'Margaret Simey: Social reformer and local councillor who challenged the Merseyside Police over the Toxteth Riots' *The Guardian* 29 July 2004 available at http://www.theguardian.com/news/2004/jul/29/guardianobituaries.politics accessed 17 April 2016

Courtman S (2007) 'Culture is ordinary: The legacy of the Scottie Road and Liverpool 8 writers' in Jones D and Murphy M (eds) *Writing Liverpool: Essays and Interviews*. Liverpool University Press: Liverpool, pp 194–209. doi:10.5949/upo9781846314476.015.

Couch C (2003) *City of Change and Challenge: Urban Planning and Regeneration in Liverpool*. Ashgate: Aldershot. doi:10.4324/9781315197265.

Czarniawska B (2002) *A Tale of Three Cities: Or the Glocalization of City Management*. OUP: Oxford. doi:10.1093/acprof:oso/9780199252718.001.0001.

Davis C (2002) 'Derelict site becomes model of inner city tranquillity' *The Guardian*, 23 January 2002. Available at http://www.theguardian.com/society/2002/jan/23/urbanregeneration. Accessed 9th April 2016.

Denington E (1972) 'New towns for whom?' in Evans H (ed) *New Towns: The British Experience*. Charles Knight & Co: London, pp 142–149.

Diamond D (1972) 'New towns in their regional context' in Evans H (ed) *New Towns: The British Experience*. Charles Knight & Co: London, pp 54–65.

Dittmer J (2010) *Popular Culture and Geopolitics*. Rowman & Littlefield: Blue Ridge Summit.

Dow C (1998) *Major recessions: Britain and the world, 1925–1995*. Oxford University Press: Oxford.

Evans H (1972) 'Appendix: Statistics' in Evans H (ed) *New Towns: The British Experience*. Charles Knight & Co: London, pp 173–192.

Fazey C (1988) *The Evaluation of Liverpool Drug Dependency Clinic: The First Two Years 1985–1987*. Research Evaluation and Data Analysis: Liverpool.

Foord M, Palmer J and Simpson D (1998) *Bricks without Mortar: 30 Years of Single Homelessness*. Crisis: London.

Friend A and Mitchell A (1981) *Slump City: The Politics of Mass Unemployment*. Pluto: London.

Freire P (1972) *Pedagogy of the Oppressed*. Penguin: Harmondsworth.

Frost D and North P (2013) *Militant Liverpool: A City on the Edge*. Liverpool University Press: Liverpool. doi:10.5949/liverpool/9781846318634.001.0001.

Grady H (2014) 'The English city that wanted to "break away" from the UK' *BBC News Magazine*. Available at http://www.bbc.co.uk/news/magazine-29953611. Accessed 8 April 2016.

Hake B (1975) *Education and Social Emancipation: Some Implications for General Secondary Education towards the Year 2000*. Martinus Nijhoff: The Hague.

Hayes G (1987) *Past Trends and Future Prospects: Urban Change in Liverpool 1961–2001*. Liverpool City Council: Liverpool.

Heseltine M and Leahy T (2011) *Rebalancing Britain: Policy or Slogan? Liverpool City Region – Building on Its Strengths: An Independent Report*. Department for Business, Education and Skills: London.

Lane T (1997) *Liverpool: City of the Sea*. Liverpool University Press: Liverpool.

Lawless P (1989) *Britain's Inner Cities*, 2nd ed. Paul Chapman Publishing: London.

Lawrence S and Pipini M (2016) 'Violence' in Cashmore E and Dixon K (eds) *Studying Football*. Routledge: Abingdon, pp 11–29. doi:10.4324/9781315737072.

McKendrick E (2003) *Contract Law: Text, Cases and Materials*. Oxford University Press: Oxford.

Meegan R (1989) 'Paradise postponed: The growth and decline of Merseyside's outer estates' in Cooke P (ed) *Localities: The Changing Face of Urban Britain*. Unwin Hyman: London, pp 198–234.

Meegan R (1990) 'Merseyside in crisis and conflict' in Harlow E, Pickavance C and Urry J (eds) *Place, Policy and Politics: Do Localities Matter?* Unwin Hyman: London, pp 87–107.

Meegan R and Mitchell A (2001) '"It's not community round here, it's neighbourhood": Neighbourhood change and cohesion in urban regeneration policies' *Urban Studies* 38 (12) pp 2167–2194. doi:10.1080/00420980120087117.

Meegan R (2003) 'Urban regeneration, politics and social cohesion: The Liverpool case' in Munck R (ed) *Reinventing the City: Liverpool in Comparative Perspective*. Liverpool University Press: Liverpool, pp 53–79. doi:10.5949/liverpool/9780853237976.003.0004.

Millington B and Nelson R (1986) *'Boys from the Blackstuff': The Making of TV Drama*. Comedia: London.

Muchnick D (1970) *Urban Renewal in Liverpool*. The Social Administration Research Trust: Birkenhead.

Murden J (2006) 'City of change and challenge: Liverpool since 1945' in Belchem J (ed) *Liverpool 800: Culture, Character and History*. Liverpool University Press: Liverpool, pp 393–485.

Of Time and the City [DVD] (2008) Terence Davies dir UK: Digital Departures, Hurricane Films.

Parkinson M (1985) *Liverpool on the Brink: One City's Struggle against Government Cuts*. Policy Journals: Hermitage.

Pearson G (1991) 'Drug-control policies in Britain' *Crime and Justice* 14 pp 167–227.

Pearson G and Gilman M (1994) 'Local and regional variations in drug misuse: The British heroin epidemic of the 1980s' in Strang J and Gossop M (eds) *Heroin Addiction and Drug Policy: The British System*. OUP: Oxford, pp 102–120.

Pickett K and Boulton D (1974) *Migration and Social Adjustment: Kirkby and Maghull*. Liverpool University Press: Liverpool.

Pinto L (2000) 'A militant sociology: The political commitment of Pierre Bourdieu' in Fowler B (ed) *Reading Bourdieu on Society and Culture*. Blackwell: Oxford, pp 88–104.

Pooley C (2006) 'Living in Liverpool: The modern city' in Belchem J (ed) *Liverpool 800: Culture, Character and History*. Liverpool University Press: Liverpool, pp 171–256.

Pooley C and Irish S (1984) *The Development of Corporation Housing in Liverpool 1869–1945*. Centre for North West Regional Studies: Lancaster.

Roberts L (2010) 'World in one city: Surrealist geography and time-space compression in Alex Cox's Liverpool' *Tourism and Visual Culture* 1 pp 200–215. doi:10.1079/9781845936099.0200.

Robson B (1988) *Those Inner Cities: Reconciling the Social and Economic Aims of Urban Policy*. Clarendon Press: Oxford.

Rogers K (2010) *The Lost Tribe of Everton and Scottie Road*, Kindle ed.Trinity Mirror Media: Liverpool.

Rogers K (2012) *Lost Tribe: The People's Memories*. Trinity Mirror Media: Croydon.

Sharples J (2004) *Liverpool*. Yale University Press: New Haven.

Sinclair D (2014) *Liverpool in the 1980s*. Amberley Publishing: Stroud.

Tallon A (2013) *Urban Regeneration in the UK*, 2nd ed. Routledge: Abingdon.

Taylor D (2009) *Liverpool: Regeneration of a City Centre*. BDP: Manchester.

Thornton P (2003) *Casuals: Football, Fighting and Fashion: The Story of a Terrace Cult*. Milo Books: Preston.

Thornton P (2013) 'Smack is back in Liverpool – But at least the approach to treating drug addiction has evolved' *The Independent*, 20 June 2013. Available at http://www.independent.co.uk/voices/comment/smack-is-back-in-liverpool-but-at-least-the-approach-to-treating-drug-addiction-has-evolved-8665586.html. Accessed 15 April 2016.

Topping P and Smith G (1977) *Government against Poverty? Liverpool Community Development Project 1970–75*. Social Evaluation Unit: Oxford.

Tulloch A (2011) *The Story of Liverpool*. The History Press: Stroud.

White I (2015) *Environmental Planning in Context*. Palgrave: London. doi:10.1007/978-1-137-31566-3.

Wilson H and Womersley J (1977) *Change or Decay: Final Report of the Liverpool Inner Area Study*. HMSO: London.

5 Back in time for the future

'Backward-facing' regeneration in Liverpool from 1980

Thus far, we have conceptualised the many meanings of 'home', and looked at home in the Liverpool context, and how it has been influenced by various processes, up to the end of what I have termed the 'forward-facing' regenerative period; a timespan within which the city displayed hopefulness, positivity, and optimism for the future. This optimism manifested itself clearly in the plans for the built environment of the city, and indeed the roads, estates, and housing that were created; although arguably hopes and plans for the future were not matched with organisation, forethought, or consideration of the needs of the people who inhabit the city (Muchnick, 1970). This chapter picks up the 'story' of Liverpool's regeneration and home trajectory at the crucial turning point when, conceptually, Liverpool can be understood as doing an about-face; giving up on the last vestiges of futuristic regeneration in favour of looking to the past – 'mining' the city's history and heritage to benefit the future. The chapter begins with arguably one of Liverpool's bleakest times when the city's only hope was to turn to the past for possible solutions.

1980s: Continued decline, political crisis, entrenchment of negative stereotypes, and the discovery of 'heritage'

By the early 1980s, the devastation started by the decline in industry dating back to the 1960s was virtually complete (Meegan, 1989). Port-related industry became well-nigh obsolete (Tulloch, 2011), the River Mersey was written off as "an open sewer" (Heseltine and Leahy, 2011, p7) and Liverpool's identity as a post-industrial city became firmly entrenched (Roberts, 2010). The social unrest of the summer of 1981, fuelled by a perfect storm of inveterate and widespread long-term unemployment, racial tension, and overzealous policing (Frost and Phillips, 2011), resulted in £11 million worth of damage and the loss of 140 buildings (Aughton, 2008) and led to tear gas being used against 'rioters' for the first time on the British mainland (Tulloch, 2011). Whilst Margaret Simey (cited in Furmedge, 2008, p93) argued that "only apathetic fools" would not have taken to the streets in the face of such duress, the majority of mainstream commentators and the population of the UK alike saw the events of 1981 as conclusive evidence of the inherent violence,[1] criminality, and disrespect for authority of a workshy populace. This is arguably best capsuled in this oft-cited opinion piece printed in the *Daily Mirror*: "They should build a fence around [Liverpool] and charge admission. For sadly it has become a 'show-case' for everything that has gone wrong in Britain's major cities" (cited in Coleman, 2004, p105).

Governmental documents released under the '30-year rule'[2] in December 2011 reveal that the preferred solution to the "intractable problem" (Rodwell, 2014, p21) of Liverpool was a 'managed decline', whereby the city and its inhabitants would be left to their fate under the mistrustful, malevolent surveillance of central government (Murden, 2006; Furmedge, 2008). The firm belief of senior government figures, including Geoffrey Howe and Keith Joseph, was that the difficulties faced by the people of Liverpool were self-inflicted, and that efforts to increase funding for redevelopment in the area would simply be a waste of money; rather, it was preferable to 'evacuate' the city of its more able/ambitious/compliant residents by encouraging relocation to more prosperous areas nearby (Travis, 2011). Further, such a policy was to be implemented as far as possible without attracting notice:

> Isn't this [financial investment in Liverpool] pumping water uphill? Should we go rather for 'managed decline'? This is not a term for use, even privately. It is much too negative, when it must imply a sustained effort to absorb Liverpool manpower [sic] elsewhere – for example in nearby towns of which some are developing quite promisingly.
>
> (Chancellor of the Exchequer Geoffrey Howe, 1981, cited in Travis, 2011, no page given)

The notion of a "concentration of hopelessness" (ibid.) in Liverpool at that time, although not the result of collective self-pity as the cabinet papers indicate, was real, far reaching, and significant (Meegan, 1994; Murden, 2006; Frost and North, 2013; Balderstone et al., 2014). The relentless series of factory closures and subsequent climbing rates of unemployment led to widespread economic devastation which was experienced by many on a deeply personal level; a resignation to the seemingly inevitable fate of the city and fear of the future[3] (Frost and North, 2013). Building on the groundwork laid by the 'working-class writers' that emerged during the 1970s, one playwright, Alan Bleasdale, created a series for television which both captured the culture and climate of post-industrial Liverpool and, crucially, revealed it to a wider audience (Hallam, 2007). *Boys from the Blackstuff*,[4] very much in the same 'social impact' vein as *Up the Junction*[5] (Millington and Nelson, 1986) and *Cathy Come Home*[6] (Murden, 2006) was more than just a 'slice of life' drama; it was a *message about* Liverpool life and how desperate it had become for many. Recognised as a "critique of British society under the Conservative Government of Margaret Thatcher" (Boland, 2008, p357), *Boys from the Blackstuff* made use of the post-industrial built environment of Liverpool, most notably the docks, as the backdrop for the crisis in masculinity and the crumbling of working-class identity experienced by the main characters (Millington and Nelson, 1986; Balderstone et al., 2014). Crucially, "when other apparently 'serious' television drama series were turning their backs on social reality" (Millington and Nelson, 1986, p171), the serial acted as a realist counterpoint to the grandeur and opulence of other contemporary dramas such as *Brideshead Revisited* and *A Passage to India* (ibid.). For the unemployed in Liverpool, it acted as a social mirror, a challenge to the myth of the 'Scouse scrounger' and, perhaps most importantly, a validation of the despair that had set in within working-class Liverpool communities (Allt, 2008b; Boland, 2008; Frost and North, 2013).

However, rather than suffer a 'managed decline', the city was instead taken under the wing of a new 'Minister for Merseyside' – Conservative Secretary of State for the Environment Michael Heseltine (Couch, 2003; Frost and North, 2013). Setting up the Merseyside Task Force working alongside the Merseyside Urban Development Corporation and Enterprise Zones, Heseltine sought to tackle the knotty and stubborn problem of Liverpool via a combination of fostering pride, optimism, entrepreneurialism, and a 'can-do attitude' in residents; a Victorian sense of obligation and leadership in local business owners; and, most significantly as it turns out, the idea that Liverpool had the potential for a centre of culture, leisure, and tourism[7] (Robson, 1988; Lawless, 1989; Couch, 2003; Tulloch, 2011). Early initiatives designed to spark regeneration through culture and heritage included the redevelopment of the central, listed, long defunct Albert Dock[8] (Couch, 2003), the hosting of the 1984 Tall Ships Race (ibid.), and the International Garden Festival of the same year (Robson, 1988; Lawless, 1989; Tulloch, 2011).

The Garden Festival in particular appears to encapsulate in one episode the 'false dawn' nature of many cultural regeneration initiatives in Liverpool. An idea developed in Germany as a means of job creation in reclaiming disused land (Lawless, 1989), Liverpool's Garden Festival, located alongside the derelict south docks, attracted 3.4 million visitors between May and October 1984 and was considered a resounding success (Avery, 2007). However, a lack of long-term funding plans meant that the site went into the hands of a private firm which ran it for two years before going into liquidation and the site being mothballed. As Couch (2003, p125) puts it:

> What should have been an asset to the local community has become instead a monument to the consequences of a failure to think through and plan the long-term use of the site.

Notwithstanding the foresight in the recognition of inherent value in a waterfront (Robson, 1988), Heseltine's involvement in the nascent Liverpool regeneration project of the early 1980s proved for many to be little more than a "massive makeover exercise" (Tulloch, 2011, p129), summed up by a striking worker in 1985 thus – "Heseltine came, they planted a few trees amongst the rubble, and went away again" (cited in Frost and North, 2013, p25).

Whilst the Conservative Michael Heseltine was attempting, from a central government perspective, to shift Liverpool towards the political right via public/

private collaboration, the political picture at the local government level was very different indeed (Parkinson, 1985; Frost and North, 2013). Having emerged and gained momentum as the 1970s progressed (Coleman, 2004), the Militant tendency within the Labour Party was, by the mid-1980s, the dominant force within local party politics. With a certain swagger which appeared to both mock and goad Westminster (Parkinson, 1985), Liverpool City Council under *de facto* Militant leadership embarked on a process of "municipalisation" (ibid., p178) which, from a fundamentalist Marxist point of view (Frost and North, 2013), sought to re-establish the local authority as chief provider of services from housing to leisure (ibid.). Militant Labour in Liverpool was subject to a series of criticisms and accusations including financial irregularities, intimidation of and aggression towards dissenters, prioritisation of the dynamics of social class to the detriment of other determining contexts (race, gender and so on) facilitated by 'workerism',[9] and ultimately deterring private investment (ibid.). Media attention paid to the 'loony left' of Liverpool compounded the already negative ideas attached to the city in the broader popular imagination (Couch, 2003; Allt, 2008c), and the legacy of the Militant era is perceived by many as atrocious (Frost and North, 2013).

However, one Militant legacy that is seen as positive by many is its contribution to the city's housing stock (Lawless, 1989; Murden, 2006; Frost and North, 2013). Recognising the plight of those in very poor housing conditions, either because they were overlooked during the radical rehousing processes of the 1960s and 1970s, or because they found themselves in a worse position as a result of the same processes, Liverpool City Council under Militant embarked on a strategy of demolishing high rises and replacing them with more traditional houses complete with front and back gardens. Totalling over 5,000, "Hatton's[10] Houses" (Murden, 2006, p456) are viewed by some as the greatest legacy of the Militant period, a promise kept and a material resource which "they couldn't knock … down after we'd gone" (former councillor Paul Lafferty, cited in Frost and North, 2013, p196). For others, however, these houses were then and are still an impediment to the continuing capitalist project in Liverpool, both as a deterrent to private house building in their construction (Lawless, 1989), and, particularly on the outskirts of the city centre, a waste of prime real estate:

> Too often, land utilisation has been sub-optimal … land that could have been used for business growth [has been] dedicated to low density housing development or other non-productive uses.
>
> (Heseltine and Leahy, 2011, p39)

Many of these houses are very close to the city centre on sites where tenement blocks used to stand, and their householders are often former tenement residents. Militant in Liverpool recognised that people quite naturally wanted to stay in their district of origin in spite of changes to working patterns and transport systems, and allowed many to do so – regardless of land value or accusations of inappropriate land use (Waddington, 2013). Further, Militant's investment in and commitment to social/municipal housing was wholly at odds with the Thatcherite 'right to buy'[11] policy of selling off council properties at cut rate prices to existing tenants (Burnett, 1986; Evans, 2003).

Arguably, this episode in Liverpool's history, coinciding as it did with both the Heysel Stadium disaster[12] and the Hillsborough disaster,[13] consolidated and settled once and for all the negative reputation of the city and the negative stereotypes attached to 'Scousers' (Millington and Nelson, 1986; Murden, 2006; Bennett, 2007; Courtman, 2007; Allt, 2008a, 2008b; Boland, 2008, 2010b; Furmedge, 2008; Devaney, 2011; Frost and North, 2013). Ideas about and characteristics attributed to Liverpool and Liverpudlians derive from several sources, most notably the news media (Boland, 2008, 2010b); however, stereotypes and reputations about place can also derive from fiction, particularly television and film (Millington and Nelson, 1986). Although several television programmes had depicted Liverpool as a place and Liverpudlian characters, some of whom were memorable for particular characteristics, arguably, it is not until the 1980s that characters emerged that embodied, in the popular imagination, characteristics viewed as attributable to *all or most* Liverpudlians. So, the fictional town of Newtown in *Z-Cars*, whilst set on the outskirts of Liverpool (Du Noyer, 2002), could have been an overspill town on the outskirts of any city; Alf Garnett's "Scouse git" son-in-law in *Till Death Do Us Part* was bolshie, but not particularly because of his place of birth (Allt, 2008a, p32); and the young female leads of *The Liver Birds*, whilst stereotypically ditsy and unlucky in love (Irwin, 2016), did not have a particularly 'Scouse' quality attached to their femininity. From the 1980s, however, the impact of Liverpudlian characters, and the idea that they are representative, appears to have gained traction.

Millington and Nelson (1986, pp169–170) discuss Alan Bleasdale's own understanding of the characters he developed for *Boys from the Blackstuff*. For Bleasdale, George Malone "stood for the traditional socialist, a trade union activist but Fabian in outlook"; Chrissie Todd meanwhile is the serial's central character as he represents the 'common man': "everything's gonna be ok, until forces that he has no control over take him over, and he's shipwrecked; his points of reference are lost". However, the best remembered character in the popular imagination is arguably Yosser Hughes, a character who has clearly developed significant mental health issues as a result of loss of employment, marital breakdown, and the strain of the search for work in order to provide for his children. Thus, rather than the old time socialist, or the average family man becoming emblematic of early 1980s Liverpool, it is the desperate, anguished, shell of a man with the "gizza job" (ibid., p168) catchphrase who comes to represent the city.

Negative imagery with specific relevance to unemployment, idleness, criminality, and reliance on state benefits further developed throughout the 1980s; soap opera *Brookside* depicted youth unemployment, anti-social behaviour, and *ennui* (Hobson, 2008), *Bread* continued in a similar vein but with the addition of cash-in-hand trading and benefit fraud, and by the early 1990s *Harry Enfield's Television Programme* had parodied the entire stereotype in *The Scousers* (Allt, 2008a). Boland (2008) notes that much of the stereotyping and caricaturing of Liverpudlians stems from the widely accepted notion that people from Liverpool are inherently funny; that there is an identifiable 'Scouse' sense of humour, often attributed to the need to develop humour as a coping mechanism in the face of adversity. This sense of humour and ability to laugh at one's own misfortune is very often accepted as a common good (Meegan, 1990, 2003; Belchem and MacRaild, 2006; Murden, 2006); however, the focus on Liverpool's seemingly universal ability to see the funny side detracts from the very serious nature of the many problems, difficulties, and traumas that the area has suffered. *Boys from the Blackstuff* is one of the few Liverpool-set pieces which challenges this viewpoint and exposes it as a hindrance rather than a help: When Chrissie's wife Angie discovers a hole in one of her shoes, Chrissie jokes that she should walk on one leg – to which she replies "It's not funny, it's not friggin' funny. I've had enough of that – if you don't laugh you'll cry – I've heard it for years – this stupid soddin' city's full of it" (Millington and Nelson, 1986, p172). Thus, *Boys from the Blackstuff* serves not only to shine a light on social issues, but also to challenge "any sentimental view of Liverpool as a city of endlessly resilient jokers who always come up smiling" (ibid.).

1990s: Consolidation of negativity, social exclusion discourse, and nascent gentrification

If the 1980s sowed the seeds of regeneration in Liverpool, against a backdrop of a worsening public image, it is fair to say that the 1990s continued in a similar vein (Couch, 2003). The global recession of the early 1990s impacted on the majority of Britain, and in particular the post-industrial cities of the north (Dow, 1998). In Liverpool and surrounding areas, this manifested itself in entrenched long-term unemployment (Marren, 2016), a bitter council workers' strike (Ironside and Seifert, 2000), thousands unable to pay the newly established Community Charge or 'poll tax' (Butler et al., 1994), endemic, sustained heroin addiction (Morgan, 2014), associated gang activity (Lavorgna et al., 2013) and, in some of the peripheral overspill housing estates, empty properties that no one wanted to occupy (Holmes, 2003). Continuing negativity and stigma relating to the Hillsborough disaster (Boland, 2008) was compounded by the abduction and murder of two-year-old James Bulger by two ten-year-old boys in a Bootle shopping centre (Davis and Bourhill, 1997), prompting Maggie O'Kane to lament "Heysel, Hillsborough and now this" in *The Guardian* (cited in Scraton, 2009, p251). The linking of Heysel – an example of football violence, Hillsborough – a state crime, and the Bulger case – a horrific, tragic and almost unique murder, on account of their shared relationship to Liverpool as a city, consolidated the notion of Liverpool as an inherently criminal and immoral place. The high profile and widely publicised dock workers' strike between 1995 and 1998 was one of the longest running industrial disputes in UK history (Mah, 2014), and strengthened the viewpoint that the city was workshy and truculent on top of depraved and violent.

Ironically, recognition of the problems impacting on Liverpool, and some level of remedy, came in the 1990s from the cause of many of them – the European

Union (Boland, 1999, 2000; Evans, 2002; Gripaios and Bishop, 2006). Identified as an area in need due to the failure of the "neo-liberal supply-side panacea … to cure the cancer of unemployment, poverty and exclusion" (Boland, 2000, p213), Merseyside qualified for Objective 1[14] funding from Europe, the first to do so in the UK (ibid.). Merseyside's Objective 1 funding programme priorities were identified as developing business, people, locations, and communities and "cross cutting themes" were identified including social inclusion and environmental issues, and a total of 1,300 projects were supported (Gripaios and Bishop, 2006). Positive results included investment in the region's higher education sector, a strengthened business sector, and the development of such infrastructural assets as the port and the airport (Evans, 2002).

The arrival and 'kicking in' period of Objective 1 funding[15] coincided with several grass-roots developments which had the impact of bringing revenue to, increasing the 'brand' of, and, crucially as it turned out, showcasing the *culture* of Liverpool. The first Mathew Street Festival, a free outdoor music event in the city's Cavern Quarter, was held during the August bank holiday weekend of 1993, and attracted 20,000 spectators (Wright, 2015). Meanwhile, following nascent attempts at venues like Quadrant Park and 051, James Barton opened Cream, a night club which sought to reject the seedy, dangerous image of the old 'clubland', and fill a void in Liverpool's music scene with dance music. Cream would very quickly establish a reputation for a "Scouse house" (Du Noyer, 2002, p221) scene which would attract coach loads of revellers from around the country (ibid.). Thus, the long-standing local recognition of Liverpool as a "party town" (Murden, 2006, p479) began to be disseminated further afield.

Heseltine and Leahy (2011) note that periods of regeneration tend to come in waves. Arguably, events such as the Mathew Street Festival, and the birth of the Cream franchise, together with the push and support of European funding, mark the starting point of a wave of regeneration which paved the way for contemporary regenerative processes, cultural revival, and event-driven regeneration (Couch, 2003; Coleman, 2004; Boland, 2013). Harms caused by previous regenerative attempts, from the destruction of communities (Couch, 2003) to "bird shit architecture"[16] (Rodwell, 2014, p28) planned and built with little or no consideration for aesthetics, retaining a certain look or feel, or lines of sight towards existing grand architecture or the river (ibid.), were acknowledged, alongside a developing concern for conservation. City centre living, which had long been frowned upon as both backward looking and a signifier of poverty, was now back on the agenda as part of a process of re-gentrification which recognised the grand Georgian and Victorian architecture of the city centre and its immediate environs as valuable assets for an emerging entrepreneurial and creative class (Couch, 2003; Coleman, 2004; Boland, 2007, 2013; Aughton, 2008; Tallon, 2013). "Heritage tourism" (Timothy, 2011), which acknowledges and celebrates buildings, artefacts, traditions, and "famous children" (Furmedge, 2008, p89), was identified as a central and vital feature of the "mega regeneration" (Tallon, 2013, p197) Liverpool needed. The crucial element of heritage tourism in the Liverpool case, the driving force that would set the city apart from other places blighted by post-industrial decline, was identified as 'culture' (Coleman, 2004; Aughton, 2008; Tallon, 2013).

Thus, regeneration initiatives that emerged during the 1990s took various forms, ranging from cultural (for example, the opening of the Museum of Liverpool Life in 1993 (Moore, 1997); the Rope Walks project[17] (Bayley, 2010)), to retail (for example, the development of the Queens Square complex (Boland, 1996)), transport (for example, the overhaul of the Gyratory and Paradise Street bus stations)[18] and public safety/surveillance (the wholesale installation of closed circuit television cameras (CCTV) across the city centre in the wake of the abduction and murder of James Bulger (Coleman, 2004)). This occurred within the context of a conscious decision made by Liverpool City Council: "We identified the city centre as our major area of opportunity" (Councillor Richard Kemp, cited in Boland, 2013, p258). What all of these disparate developments shared, though, was a focus on consumption and a commitment to private investment (Couch, 2003; Coleman, 2004; Tallon, 2013).

Coleman (2004, p66) identifies this period in Liverpool's history as characterised by an emergent "neoliberal statecraft" which facilitates the furtherance of a neoliberal agenda through 'partnership' – essentially a marriage between the democratically elected local authority and private enterprise, or a "shift in civic leadership from municipal socialism to urban entrepreneurialism" (Boland, 2013, p255). The influence of corporate bodies over decision-making processes ensures that wealth accumulation is prioritised over other interests, such as access to public space and protection of civil liberties. Thus, in the Liverpool case, this period marks the start of a collusion between the local authority, the police, and the private sector; "partnerships [who] engage in constructing a form of official discourse through which a rationale and legitimacy of rule can be forged along with the construction of the objects of partnership power" (Coleman, 2004, p121). Such "mega regeneration" (Tallon, 2013, p197), which has retail and leisure at its very core, must operate within a climate that allows it to flourish; an environment which must "police and monitor the 'debris' of neoliberal urban visions – litter, graffiti, the homeless and prohibited street trading" (Coleman, 2004, p70). By the end of the decade a dedicated economic development company, Liverpool Vision, had been created to both attract and facilitate greater public-private collaboration in the Merseyside area (Couch, 2003).

Lovering (2007, p357) notes that: "cities are the victims, dupes, recipients, or targets of neoliberalism. They have it done to them, rather than themselves helping to bring it about". Liverpool's neoliberal statecraft project, then, relies on a positive public image for the city and its inhabitants in order for it to succeed. For Liverpool Vision, one means of securing this was to create an Academy, attributed with the moniker The Charm School, to promote civility, politeness, and manners amongst city centre service staff (Coleman, 2004). Another measure adopted by Merseyside Police, in conjunction with Liverpool City Council, was to engage in 'quality of life', 'zero tolerance'[19] approaches to public space management, resulting in "unattractive and irksome reminders of social inequality…being airbrushed out of the imaginary representations and material realities of city life" (ibid., p81).

Similarly, the city's Business Improvement Districts (BIDS), which take levies from members to 'improve' the areas where they operate (Peel et al., 2007), have a particular focus on safety, and fund both two police officers and a team of "Evening ambassadors…to report and address anti-social behaviour, cleansing and environmental issues" (Liverpool BID Company, no date given). Coinciding as it did with New Labour's identification of and concern with social exclusion (Levitas, 1998), "this process contrives to appeal to a 'new' constituency in Liverpool: an upwardly mobile, prosperous and entrepreneurial citizenry" (Coleman, 2004, p146).

In order to ensure the compliance of the populace for this renaissance, and to promote an image of a prosperous, dynamic, inclusive Liverpool, other initiatives were also required to address the inequalities and social injustice that persisted within the city (Meegan and Mitchell, 2001; Heseltine and Leahy, 2011; Tallon, 2013; Ilan, 2015). Power and Wilson (2000) note that, when central government produced an index of local deprivation in 1998 detailing the most deprived areas in England, Liverpool was ranked at number one whilst Knowsley, bordering Liverpool and containing several overspill developments, was ranked at number nine. Thus, the area continued to bear the hallmarks of New Labour-defined social exclusion,

> a shorthand label for what can happen when individuals or areas suffer from a combination of linked problems such as unemployment, poor skills, low incomes, poor housing, high crime environments, bad health and family breakdown.
>
> (Social Exclusion Unit, 1997, cited in Levitas, 2006, p125)

whilst at the same time aspiring to "world class status"[20] (Liverpool City Council, 2008; cited in Kinsella, 2011, p241). In this sense 1990s Liverpool can be understood as a 'dual city' offering "divergent and mutually exclusive experiences to the rich and poor" (Ilan, 2015, p62).

Remedies for the varied symptoms of social exclusion, from surplus housing in undesirable 'sink estates' to long-term unemployment,[21] came in numerous forms, many of which were characterised by the now familiar blend of public and private initiatives. Sure Start centres were created to tackle child poverty, educational performance, and lack of child care as a barrier to work for parents (Shaw, 2010), 'sustainable communities'[22] approaches were taken with districts deemed to be failing in terms of housing provision (Tallon, 2013), and 'city deals' and 'new deals' like the Kensington New Deal,[23] were launched, whereby "each deal is bespoke and reflects the different needs of individual places" (ibid., p113). Peel Holdings,[24] established during the 1970s, emerged as a key private sector player with the acquisition of a majority share in Speke Airport in 1997 (Peel Property & Land, 2015).

However, as the 1990s segued into the new millennium, it became apparent that in some cases the drive to eradicate social exclusion had given rise to additional problems rather than solving existing ones. A new phenomenon, termed "regeneration led dereliction" by Furmedge (2008, p83), developed, as remedies for the effects of social exclusion left some areas of the city in a worse state than previously. Notable examples include the Edge Lane district, just to the west of the outskirts of the city centre, where significant numbers of houses were left empty for years until plans to widen and improve Edge Lane, the main arterial route into the city centre from the M62, were confirmed and put into action (Allen, 2008); and the Boot Estate in Norris Green, which was partially demolished to facilitate the development of a new "urban village" (Mulhearn, 2003) before Liverpool City Council reneged on its pledge to fund the project (ibid.). Many houses on the estate remained empty for more than eight years, rendering the area a virtual ghost town (Cousins, 2013). This 'regeneration-led dereliction' and its negative consequences would set the tone for further regeneration-led disappointment alongside perceived regeneration successes in the following decade (Boland, 2008; Furmedge, 2008).

Millennial Liverpool: regeneration, re-gentrification, and 'everything is awesome' discourse

Boland (2007, p1027) suggests that, in the contemporary advanced capitalist, neoliberal climate, "cities are conceived as capitalist commodities that can be redesigned and marketed to consumers in a marketplace". Against this backdrop, 21st-century Liverpool continues to live in the shadow of, and remains dominated by, the accolade of European Capital of Culture (ECOC) for 2008 (Furmedge, 2008; Boland, 2010a; Devaney, 2011; Cox and O'Brien, 2012). Devaney (2011, p172) conceptualises 2008 as a watershed moment, marking the shift from one epoch to another:

> Here is a Liverpool where thirty-somethings go misty-eyed at the thought of simpler times before years had cultural themes while their children gaze in awestruck wonder at legends of gargantuan mechanical spiders and the merciless invasion of a mythological race of half-sheep, half-bananas.

However, as demonstrated, millennial Liverpool – the contemporary Liverpool characterised by heritage, culture, and ostentatious consumption, has identifiable roots in the 1980s and was fully consolidated in June 2003, when the ECOC accolade was announced. The people of Liverpool were reported to be intensely proud of this achievement, which would act as a counterbalance to, and a rebuttal of, the persistent negative reputation that the city struggled to shake off, as well as a useful source of revenue (Allt, 2008; Furmedge, 2008; Boland, 2010a, 2010b; Cox and O'Brien, 2012).

Boland (2010a) notes that Liverpool was not expected to win with their ECOC bid and Newcastle-Gateshead was the bookmakers' favourite; what ultimately swung it for Liverpool was the 'whole city' nature of the bid – everyone was behind it, and everyone would benefit from it[25] (ibid.). Whilst there were some who maintained their view that "Scousers are stupid and aggressive, and they wouldn't know culture if they fell over it" (McIntyre Brown, 2001, cited in Devaney, 2011, p174), others recognised the variety of 'traditional' elements of culture that the city had to offer, from music, sport, and art to architecture and creative industries (Roberts, 2012). Culture, however, can obviously take various forms, and the term can have different meanings. Culture can be applied to the customs and rituals of a population, or it can be employed as a synonym for climate or atmosphere – either good or bad – as in, for example, a culture of success, a culture of enterprise, a culture of negativity, a culture of bullying. The term developed from agriculture, and the notion of cultivation: The creating, nurturing, and monitoring of, for example, a crop or a sample of bacteria created in a petri dish (Bauman, 2011; Mahler, 2013; Low, 2017). Bauman (2011, p5) defines culture as "a set of preferences suggested, recommended and imposed on account of their correctness, goodness or beauty" – in other words, an ideal, something to aspire to. However, in the case of ECOC it is considered almost universally to refer to the arts (Furmedge, 2008; Higginson, 2008) and it is viewed as 'high culture'.[26] Like the concept of home, 'culture' in this sense seems to be viewed as an inherent good; when an activity, process, or object has the label 'cultural' attached to it, it is perceived to be both beneficial in and of itself and vital to creative processes and wealth accumulation (Bauman, 2011).

Once the accolade was announced, the city's "renaissance managers" (Coleman and Hancock, 2009) set about the task of 'delivering' culture in time for 2008. Furmedge (2008, p82), noting that

> since the city of Liverpool was colonized by a cadre of regeneration professionals in the wake of the 1981 Liverpool 8 uprising, the locals have grown used to a succession of suits and generic middle England accents being paraded as saviours of the inner city,

details the steps taken to provide oversight and management of the ECOC preparations. Liverpool Culture Company was created, at significant expense, some of which was raised by diverting funds earmarked for community engagement projects in the districts,[27] to coordinate preparations. Robyn Archer, an Australian artist and performer with no prior links to the city, was appointed as Artistic Director on a six-figure salary, only to be removed after spending just eight weeks in the city during two years (Herbert, 2006). Furmedge concludes that the Culture Company's failure to (a) audit what cultural assets were already in the city and (b) consult and engage the people of the city points to an arrogant body for whom "the primary purpose … became … the promotion and survival of the company" (p85).

For Furmedge, the trajectory of the Liverpool Culture Company follows an established regeneration pattern:

> A couple of 'gateway' schemes, some roadworks and a bit of CPO [compulsory purchase orders] later, all the money's gone and so are the erstwhile regenerators, off to pastures new with a CV showing how they held those militant Scousers at bay to deliver spend targets on time.
>
> (2008, p93)

Further, the statecraft-constructed blurring of the lines between the private sector and the public sector rendered public accountability for the use of funding and decisions regarding events and projects virtually impossible (Higginson, 2008). Thus, ECOC effectively became a money-making endeavour for "the regeneration professionals" (Furmedge, 2008, p82), rather than a cultural accolade which may bring with it an economic boost for the city.

Furmedge (ibid.) further notes that previous ECOC cities made use of the whole urban area; Lille, for example, had a 'cultural centre' in each of its districts. In the Liverpool case, however, ECOC barely strayed from the city centre other than for some minor, seemingly tokenistic events, for example, the temporary installation of over 100 'superlambanana'[28] replica sculptures throughout the districts (Jones, 2008). Devaney (2011, p178) synopsises the concern:

> But what about outside of the city centre? What about our neighbourhoods and communities? The council was accused of neglecting its 'bread and butter' responsibilities in its dogged pursuit of the 'culture' prize.

"Brand Liverpool" (Roberts, 2010, p202), then, very clearly prioritised the city centre, and particularly the waterfront (Higginson, 2008; Boland, 2013); in doing so, it simultaneously ignored the cultural richness in existence in the districts, and starved them of further stimulation and funding (Allt, 2008a; Furmedge, 2008). Further, as Lovering observes (2007, p362), prioritisation of city centres is constructed and presented as beneficial for all city inhabitants:

> In cities with enormously divided labour and housing markets, all can nevertheless share the virtual egalitarianism of consuming the new skyline or waterfront vista, the beautifully lit evening waterside panorama, and the glittering fetishistic world of 'frozen commodities' that is the new shopping mall.

Adopting the role of *flaneur*,[29] Boland (2010a, p633) discusses his return to his home city during 2008, and his search for evidence of ECOC within the districts:

> My travels took me through many parts of the city during 2008 and 2009, for example, the city centre and housing estates in Speke, Garston, Halewood, Netherley, Norris Green, Toxteth and Croxteth. There was no obvious evidence of cultural events taking place, no promotional signage, nor was it a conversational issue for local residents ... peripheral estates showed little, if anything, to indicate Liverpool was the ECOC. The only constant reminders were *08* logos on Merseybus services as they traversed the city's streets.

This was in stark contrast with what Boland encountered in the city centre:

> My journey exposed a life *a world away* from the stylish regeneration in the city centre. The contrast was startling with impoverished housing estates suffering unemployment, deprivation and crime against a city centre thronged with yuppie apartments, four- and five-star hotels, plush eateries and exclusive stores.
>
> (ibid., emphasis in original)

These distinctions did not go unnoticed by residents of the city living outside of its centre, who reported feeling disengaged and excluded (Furmedge, 2008; Higginson, 2008; Boland, 2010a; Taylor, 2011). For Higginson (2008, p55), this exclusion is the result of the 'arm's length' approach of the Culture Company, coupled with poor use of resources and, crucially, views about the Liverpool populace:

> The levels of local exclusion have an air of condescension and contempt linked to 'class'. We are not supposed to make our own culture. Instead we have to sit still until a cultural clique reveals what is to be enjoyed.

This conception of culture as 'new' and 'alien' to the people of Liverpool is instructive in a number of ways.[30] It prioritises a very one-dimensional understanding of culture – 'high' culture – over others, as a way of "saving the savage from his [sic] barbaric state" (Bauman, 2011, p9), whilst negating both the wide spread of cultural activity at grass-roots level, and the fact that the *whole city* nature of the initial bid was what secured the accolade in the first place (Furmedge, 2008).

Taylor (2011, p164) details the rise and fall of hopes for ECOC within the city:

> In particular the city's cultural community, which had struggled to survive through years of austerity, felt that its role was finally being recognised, [instead] the city's artistic fringe found itself being pushed out of its studios and venues by the rapidly developing legions of bars and flats.

This disappointment and disenfranchisement did, however, lead to the emergence of "a loose anti-2008 movement" (ibid.) of independent artists, retailers, and restauranteurs who sought to offer an alternative to the capitalist, neoliberal appropriation of ECOC and became a "rallying point of the 'independent' and 'local' against the 'corporate' and 'global'" (ibid.). Liverpool's resilience and resistance to the wholesale consumption of the city's cultural elements by multi-national corporations is, for Taylor (ibid., p165),

> Liverpool's greatest contribution to the wider world for having been awarded ECOC: to have been the place that questioned, even deconstructed, the whole concept, in the process changing the way many people think globally about concepts of culture, cities and regeneration.

Further, as Devaney (2011, p179) notes, this challenge to the corporatisation of Liverpool during ECOC is indicative of both the grass-roots culture, and the stubborn, 'bolshie' nature of its populace:

> Refusing to sign up to 'the Corpy' version of culture … a counter-Capital of Culture was born, a celebration of a 'warts and all' Liverpool that refused to hide its dereliction. This was real Scouse culture. And this was radical.

After Liverpool's ECOC year was concluded, work started in earnest to measure the success of the year in terms of impact, legacy, and the extent to which the city had been 'regenerated'. Liverpool as ECOC is notable in that it was the first to have a longitudinal research project developed alongside it, to monitor the implementation of what has come to be called "The Liverpool Model" of culture-led regeneration (Cox and O'Brien, 2012). Impacts 08: European Capital of Culture Research Programme published their final report in 2010, and, whilst acknowledging that the year had been a success as a cultural event, the report authors also noted that both the bid itself and early preparations made unrealistic projections about the regenerative potential of the accolade, and more should have been done to manage expectations (Garcia et al., 2010). Thus, whilst some commentators, for example Foley et al. (2012, p5), concluded that

> Liverpool 08 was deemed a success for the way in which it engaged locally with the culturally 'uninitiated' whilst generating positive media exposure nationally and internationally for its 'signature' cultural events, exhibitions and artworks,

there was recognition elsewhere that the Liverpool model did not live up to its own hype. Cox and O'Brien (2012) posit that the notion of a generic cultural regeneration model, that can be applied to any urban cultural event, is a falsehood; rather, a unique confluence of phenomena specific to Liverpool at that point in time made the event successful. Connolly (2013) goes further, arguing that the bid, influenced by New Labour cultural policy which emphasised both public/private collaboration and culture as a cure-all for urban blight, made the mistake of conflating cultural activity with solid economic and social policy. In doing so, it consolidated the assumption that socio-economic deprivation equates to a cultural void, ignoring "the fact that one might be on income support, captain the local football team on a Saturday and play the clarinet of a Wednesday night" (ibid., p175).

Thus, Furmedge's (2008, p96) concern that "Liverpool 08 [is] going to be the fur coat and no knickers Capital of Culture" appears to have been borne out, at least in part. Whilst the event might have delivered in terms of cultural celebration, in spite of its prioritisation of 'high' culture and failure to acknowledge and engage many forms of cultural activity already in existence, it was unable to keep all its promises in terms of delivering the city from socio-economic deprivation (Garcia et al., 2010; Cox and O'Brien, 2012; Connolly, 2013). However, since 2008 Liverpool has witnessed significant change – to the built environment of the city centre, to the way the people of Liverpool 'do' culture, and to the lives of Liverpool people themselves. It is to these changes that we will now turn.

Bauman (2011, p16) contends that

> the culture of liquid modernity has no 'populace' to enlighten and ennoble; it does, however, have clients to seduce. Seduction, by contrast with

> enlightenment and ennoblement, is not a one-off, once and for all task, but an open-ended activity.

Arguably, this is applicable in terms of contemporary Liverpool; the populace is less and less treated as a citizenry, and more and more as a client base who must be won over by clever marketing – a sales patter which gradually reconstructs city-wide understandings of what matters, what is valuable, what should be prioritised, and what is best for the future of Liverpool. Hegemonic, statecraft-endorsed decisions about how best to 'sell' the city are clearly visible when examining the evolving built environment of the city centre.

Roberts (2010, p202) suggests that Liverpool's cityscape has a 'legible image' which, with its distinctive architecture including two visually striking cathedrals and St John's Beacon tower, is instantly recognisable and instantly readable. The iconic waterfront was declared by UNESCO to be a world heritage site in 2004 (Murden, 2006; Devaney, 2011), and its three main attractions – the Liver Building, the Cunard Building, and the Port of Liverpool Building – were, in the run up to ECOC, attributed a new moniker: "A group of bigwigs on a lunchtime stroll decides that the Cunard, Liver and Port of Liverpool buildings were to be referred to as the 'Three Graces'" (Furmedge, 2008, p91). This labelling of these buildings as "spaces of cultural consumption" (Boland, 2007, p1027) affords them additional gravitas and further facilitates unique place identity.

Liverpool's past having already been established as Liverpool's future, the ECOC announcement encouraged both greater private/public sector funding and the impetus to enhance and further 'heritagise' the waterfront and its immediate environs (Taylor, 2009; Boland, 2010a; Heseltine and Leahy, 2011; Tallon, 2013). Devaney (2011) notes that preparations ahead of the city's 750th anniversary in 1957 included the hasty and lamentable decision to relocate Great Homer Street Market and demolish the city's much-loved Overhead Railway. However, a combination of slow post-Second World War development and "fortuitous neglect" (Taylor, 2009, p134) meant that much of the detritus of the industrial and seafaring heyday of the city was still intact, and could be utilised to maximum potential in the remaking/remodelling of Liverpool for the future.

Centring Liverpool One

Arguably the centrepiece of 'millennial' Liverpool is Liverpool One, the "mall without walls" (Carter, 2008, cited in Kinsella, 2011, p245) which is the end product of the Paradise Street Redevelopment Area on land owned by the estate of the Duke of Westminster (Taylor, 2009). Working on the premise that Liverpudlians are expert shoppers and had no outlet for their retail needs (ibid.), Liverpool One is "forty-two acres of straightforward, uncomplicated, capitalist pleasure" (Devaney, 2011, p178) which has something for everyone, from South John Street, a rendition of 'typical' high street, to Peter's Lane, a more select, 'high end' shopping experience (Taylor, 2009). The Liverpool One development is notable for a couple of reasons. First, the complex was carefully planned around existing historic buildings which were deemed to have merit, for example, the Bluecoat building, and sought to restore both the pre-Second World War street pattern, and lines of sight to the river, much to the annoyance of the 25 different architects involved who felt their designs had to take a back seat as a result (Jones, 2016). Second, it represents a merging of public and private space; people are free to enter the complex, and may not even realise they are doing so, yet the complex is very much privately owned and privately policed – anybody deemed 'unsuitable' can legitimately be ejected. Thus, the opening of Liverpool One brought concerns that "traditional rights of way were to be replaced by 'public realm arrangements' policed by private guards known as 'quartermasters' or 'sheriffs'" (Minton, 2012, no page given). Lack of clarity for pedestrians as to where the complex boundaries are appears to have been by design; Taylor (2009, p46) cites design team member Trevor Skelton:

> Within the scheme there is enough complexity so people can wander – they don't have to follow a prescribed retail triangle or a prescribed route. They can wander around and then they're never quite sure when they're in Liverpool One and when they're in the rest of the city.

For Taylor (ibid., p137) Liverpool One is distinctive:

> In an age when the general thrust in retail is towards homogeneity – where a high street in one part of the country can look pretty much like another, many hundreds of miles away, it is heartening that the idea of 'place' can still be a driver in a large, city centre urban regeneration scheme.

Thus, because the complex has had to be fitted around existing buildings deemed to have historic value, and work with them in a sympathetic manner, it has avoided falling into the trap of becoming a clone of other, less well-thought-out shopping areas (ibid.). However, the retail and leisure scene in contemporary Liverpool *is* reflective of other cities across the country and across Europe. Evans (2003) and Tallon (2013) both locate Liverpool against a backdrop of similar homogenised towns, as a result of the rise of global brands (Czarniawska, 2002; Lovering, 2007), and both Murden (2006) and Boland (2010a) refer to it as an 'identikit city' – generic, as if built with uniform parts and from an instruction manual common to all retail areas. Rather than being inherently distinctive in terms of its built environment, Roberts (2012, p205) suggests that the opposite is true in that, whilst Liverpool is frequently used as a film set, it

> all too rarely 'plays itself'; that is, the productions are not place-specific to the extent that landmarks and locations are expressive of the city's on-screen identity and urban character. Not so much the 'star' as 'body double' … Liverpool's cinematic geographies have served as a stand-in for, amongst others, Cannes, Vienna, Moscow, St Petersburg, Dublin, Amsterdam, Rome, New York, Chicago, Paris, London and war-time Germany.

Therefore, the capitalist processes which developed the city in the 19th century, and the capitalist processes which are pushing a "retail-led regeneration" (Tallon, 2013, p203) in the 21st century, are rendering Liverpool a bland 'any city' rather than resulting in distinction (ibid.). In this sense, the drive towards best practice and the establishment of the 'hallmarks' of regeneration success have led to "'copycat' urban design strategies and 'karaoke architecture'" (Boland, 2008, p1028)

Taylor (2009) further suggests that Liverpool One is the real, true legacy of ECOC and, arguably, he has a point. Tallon (2013, p231) posits that "culture has become commodified as a means to an academic end … associated with product development, marketing strategies, branding and image" – given the strong presence of public/private partnership in the city, and the extent of the reach of Peel Holdings as a mega-investor (Harrison, 2014), it is fair to conclude that Liverpool is essentially a multi-national investment site first and foremost; any organic, independent cultural activity is a bonus. The prioritisation of multi-national companies over culture is perhaps best embodied in the fate of Quiggins – a quirky, alternative shopping centre in the heart of what is now Liverpool One, which was "forced out of its city centre building by local planners and Grosvenor" (Boland, 2010a, p635) and has struggled as a brand to find a suitable home and a post-move identity since (ibid.).

For many academics and commentators, the physical regeneration of the city centre in preparation for, or as a result of, ECOC year, is little more than window dressing. Devaney (2011, p179) refers to it as a "whitewash" whilst Allt (2008a, pp7–8) views the process as nothing more than

> applying a coat of Dulux gloss over the city centre … below that thin coat of Dulux there are thousands of unheard souls just waiting to have a bit of a dabble at this culture thing – if given the chance.

Liverpool city centre, then, after spending "four years as a pile of rubble surrounded by puddles" (Januszczak, 2009, cited in Kinsella, 2011, p240) belies both the level of inclusion within the ECOC cultural project and the level of poverty, social malaise, and social exclusion persisting around the city's districts (Boland, 2007; Aughton, 2008; Coleman and Hancock, 2009; Tallon, 2013). To put it another way, the ordinary people of Liverpool, who live and work within the city and its environs, have not necessarily seen the benefits of the culture- and retail-led regeneration that has taken root in the last few years.

Fagan (2004) notes that high-rise living, i.e., in multi-storey buildings, comes in and out of fashion, just as areas of cities fluctuate in terms of their popularity. Areas like, for example, Abercromby[31] – virtually abandoned and left to ruin by the 1980s – are now celebrated for their Georgian architecture and are once again desirable places to live (Speake and Fox, 2006), and there is an identifiable pattern of dockside warehouses being turned into 'executive' apartments for "loft-living" (Evans, 2003, p25). Meanwhile, student accommodation, long the domain of university-subsidised suburban 'halls' and the terraced streets of the Smithdown

Road area (Potschulat, 2016), is now springing up in a 'luxury' format across the city centre (Evans, 2003; Aughton, 2008; Rodwell, 2014) as students are encouraged more and more to be "apprentice gentrifiers" (Potschulat, 2016). However, the fact that there are examples of high-rise executive and student accommodation sitting empty right across the city centre is an indicator both of the prohibitive cost of city centre living and a market flooded with more apartments than can be let (Rodwell, 2014).

Patterns of 21st-century housing in the inner areas, the districts, and the suburbs, by no means universally reflect the perceived prosperity of the city centre. Chris Couch had high hopes for the Kensington New Deal for Communities project when he published his book in 2003, describing it as:

> A new programme to tackle multiple deprivation in the most deprived neighbourhoods in the country in an intensive and co-ordinated manner … . It is too early to comment on the outcomes of the Kensington NDC and its impact on the area but the scale of the programme is indicative of a serious commitment to change.
>
> (p182)

However, by 2008, Kensington Regeneration – the body set up to deliver the NDC programme – had to concede that "Kensington NDC area still has some way to go in attaining socio-economic standards comparable with those across the city of Liverpool" (Kensington Regeneration, 2008). Similarly, plans to redevelop the north end of the city have come and gone (Taylor, 2009; Heseltine and Leahy, 2011; Rodwell, 2014), and there are approximately 9,000 empty houses across the city currently (Coming Home Liverpool, 2017), yet there is a perceived acute shortage of housing (Parry, 2016). Difficulties faced every day by ordinary working Liverpool families were highlighted by the Getting By? Project (2015), which uncovered insecure and short housing tenures, rising private sector rents, an acute shortage of social housing, zero-hour contracts, and fuel poverty, all of which exacerbated by a local economy increasingly dominated by the service and tourism industry, characterised by low pay. In areas deemed unpopular by Liverpool City Council and subsequently neglected and allowed to plunge into dilapidation, houses are being sold off for a sum of £1 subject to assurances of financial ability to restore the property to a liveable state (Atkinson and Maliene, 2015).

Roberts (2012, p200) starts an article about Liverpool with the following opener:

> To begin with a well-worn cliché: this is a tale of two cities. Cities that share the same geographical co-ordinates but that in other respects defy conventional processes of mapping and the semiotic inscription of place and identity.

This sentiment is very much applicable to the built environment of the city as it moves further into the 21st century. The city centre continues to develop through ongoing regenerative activity; every time one visits the city centre, one spots a newly opened shop or restaurant, a new construction site, or a newly completed building. Meanwhile, although physical change is apparent in the districts, it occurs at nowhere near the same speed, nature, or quality of that of the city centre. Contemporarily, physical regeneration happens at a much slower pace in the districts, and to limited effect. This prioritisation of physical regeneration of the city centre/waterfront over the districts is further mirrored in the cultural regeneration of the districts which, arguably, occurs on an even lesser scale.

Aughton (2008, p300) cites Sir Bob Scott, instrumental in Liverpool's successful ECOC bid, talking about the city in 2001:

> Liverpool has already chosen culture as its future. Hundreds of millions of pounds are being invested in cultural projects. Liverpool's heritage and cultural wealth is already world famous and just yesterday the city was voted favourite day out by the people of the North West. But it is the involvement and enthusiasm of our communities which will help us clinch the title.

This soundbite is instructive in a number of ways. It highlights culture as the future of the city, but acknowledges that it is Liverpool's past – or heritage – which provides that culture. Further, it stresses the role of the Liverpool community in securing this cultural future. Some years later, current Liverpool city mayor Joe Anderson (2015) expressed the view that "culture is the rocket fuel of regeneration". However, cultural regeneration in millennial Liverpool is much less clear, and much less tangible than the visible physical regeneration that the city centre has undergone.

Alongside the consensus within the regeneration context that culture equates to the arts (The Warwick Commission, 2015), it is recognised that culture takes various other forms. Bauman (2011) notes that culture is related to class divisions, in that to be elegant, sophisticated, and refined is to be of a higher class and, therefore, cultured. For Mahler (2013), culture is not necessarily something that can be seen, heard, or touched; rather, it is a feeling – a sensation of comfort that derives from being put at ease through familiarity, routine, traditions, and rituals. Courtman (2007, p194), however, draws on Raymond Williams' position that "culture is ordinary", in the sense that culture occurs at ground level and on a day-to-day basis; a form of habitus that shapes and structures every aspect of human activity and understanding. However, there is also consensus amongst academics examining culture and regeneration, in Liverpool and elsewhere, that culture also equates to consumption (Meegan, 2003; Coleman, 2004; Belchem, 2006; Boland, 2007; Bartoletti, 2010; Boland, 2010a ; Bauman, 2011; Erll, 2011; Mahler, 2013; Ilan, 2015).

The Warwick Commission was put together in 2013 to consider how culture could contribute to the development and prosperity of Britain. In its final report, the Commission made clear its view that culture clearly has some economic value, and that cultural activity can act as a source of revenue for local authorities subjected to central government austerity measures (2015). The way that this has played out in the Liverpool context is that cultural events are staged, and arguably prioritised over other forms of cultural regeneration, to the point that we can understand Liverpool's current regeneration strategy as "event-led" (Liu, 2015, p28). Contemporary cultural events in Liverpool are nearly always a reflection of the city's history – whether they be derivative of maritime heritage, for example, the International Mersey River Festival (McColgan, 2015), music heritage, for example '50 Summers of Love' (Culture Liverpool, 2017), or commemorating a particular date, for example the First World War centenary celebrations (McColgan, 2015). The Warwick Commission (2015, p66) appears to take issue with such an approach:

> The role of cultural organisations as strategic partners in the more fundamental place-shaping role, building and moulding local communities and identities, remains underdeveloped. As a result, whilst the accounts of place that civic leaders give are often redolent of local pride and distinctiveness, the economic, cultural and social strategies that are seen to comprise place shaping often lack such distinctiveness, are based on *superficial 'famous dates and people'* idea [sic] of place identity or even disregard local cultural expression entirely (emphasis added).

Reliance on "superficial 'famous dates and people'"[32] is very much redolent of the approach adopted by Liverpool City Council and its partners – in fact, Joe Anderson reportedly believes that Liverpool does not currently do enough to celebrate its status as the home city of The Beatles, and sees this as a potential further historic area to exploit (McColgan, 2015). This is, however, part of a wider trend of a "memory industry" (Erll, 2011, p3) and, more specifically, "memory tourism" (Bartoletti, 2010, p23). Drawing on Pierre Nora's conceptualisation of *lieux de memoire* as a site of collective memory, Erll (2011) recognises that the collective memory is a social construction, in that what we are encouraged to remember is delivered or revealed to us via "specialised carriers of tradition" (p29) who are authorised to establish 'official' memory as opposed to 'vernacular' memory. Thus, a process develops whereby what we remember and what we forget becomes politicised, in that what is deemed memory-worthy is dictated by the powerful. With reference to Derrida's (1995) concept of "archival fever" (p51), Erll identifies "a contradictory drive, or desire, to collect and remember and at the same time to repress, destroy and forget" (ibid.). It is possible to use Erll's analysis to explain event-led regeneration in Liverpool; Liverpool City Council and its partners act as "specialised carriers of tradition" (p29), deciding what to remember and establishing the city's waterfront, or St George's Plateau, as the designated site for remembering.

Like Mahler (2013), Bartoletti (2010) identifies culture as a 'feeling', particularly in connection with the past. Here nostalgia, which Bartoletti (2010, p24) characterises as a "typical modern illness" is exploited and commodified via "memory tourism" (ibid.). Significantly, she likens the commodification of nostalgia to sensations of homesickness; specifically, she understands nostalgia as homesickness, or *heimweh*, for a home that can no longer be returned to, as the nature and 'ways' of that former home have gone with the passage of time, and can never be located again. In this sense

> nostalgia can also be a resource, in that it represents a form of wish that can't come true – at least in its former authentic sense. While it is true that one can never really 'return home', it may instead be possible to simulate the existence of a mythical home through its invention and reconstruction. In reality, this will not be a place of memories but rather a place for consumption and entertainment: if the mythical home has disappeared forever, the marketplace can therefore continually re-invent it, creating new occasions for consumption.
>
> (Bartoletti, 2010, p25)

Arguably, then, an event-driven regeneration can continually find ways to memorialise the past and, certainly in the case of Liverpool, evoke notions of a past home that can never be recaptured, but can be celebrated via specified events at specified locations. Recognising the potential value of 'mega events', and their ability to attract worldwide attention, Liverpool City Council and its partners have ensured that, for example, the waterfront is fully equipped with "plug and play" (Kemp, 2015) technology to facilitate live broadcasting.

Lovering (2007, pp360–361) notes that regeneration is more and more characterised by performance, not just in terms of events and displays, but also in terms of roles 'performed' by invested individuals and organisations:

> Urban regeneration is regularly accompanied by a performative shift in governance. This is sometimes quite literally a matter of stage management as in the municipality-funded cultural festivals, music performances, street performers, etc., that characterise 'regenerated' areas . ... But the term performativity can also be used to refer to something more subtle and significant than conscious 'play-acting'. This second usage invokes the fact that some discourses (or 'speech acts') *can have the effect of bringing about that to which they refer* (emphasis added).

Thus, in a 'fake it till you make it' manner, the *performance* of a regenerated city, if performed enough times, will ultimately become a self-fulfilling prophecy, in that the performance will come to be understood as reality (ibid.).

To summarise, millennial Liverpool portrays itself as "a city at ease with itself – post-conflictual, socially integrated and harmonious" (Coleman, 2004, p233), where "everything is awesome" (*The Lego Movie*, 2014) for those who commit to the regeneration ideal and can shed their Liverpudlian cocoon and emerge as a Liverpolitan butterfly. Entrenched negative stereotypes and reputations persist, but they have been joined by others – the city is now established in the popular imagination as a 'party town' characterised by hedonism and ostentation (Murden, 2006; Boland, 2010b; Devaney, 2011) and as a city break destination where couples can prove their commitment to each other with a 'love lock' attached to a chain at the waterfront (Browne, 2014). Liverpool women are portrayed as glamourous, overly concerned with their appearance, and somewhat vacuous – characteristics embodied by Liv and Liz, the 'Liver Birds' developed as a marketing gimmick for Liverpool One.[33] The city views itself, and promotes itself, as a "world class city" (Liverpool City Council, 2008, cited in Kinsella, 2011, p242); a destination and a site for investment and growth, just like, for example, Barcelona, Beijing, or New York. However, the reality is that, particularly outside of the city centre, not much has changed. The districts have largely been ignored within contemporary regeneration drives, and there has been little in the way of 'trickle down'. Whilst the city centre has undoubtedly been revitalised, the process has not been inclusionary – cities have long been viewed as a site of peril and degeneracy (Graham and Clarke, 2001), and those who are constructed as embodying such characteristics are no longer welcome. Thus, begging, rough sleeping, and street drinking are 'cleansed' from the city centre along with "people carrying Netto carrier bags and young people skateboarding" (Carter, 2008, cited in Kinsella, 2011, p245) via statecraft-produced officials and the invocation of public alcohol laws (Kinsella, 2011). Arguably, then, Liverpool is in the midst of a period of revanchism (Boland, 2010a) whereby public space is being 'reclaimed' for the law-abiding citizenry who are worthy of presentation as 21st-century Liverpolitans – ambassadors who are 'on message' and will aid, rather than hinder, capitalist regenerative processes.

There are no signs of Liverpool changing direction any time soon. This is indicated, for example, in the proposed Liverpool Waters project. Boland (2013, pp259–230) describes the project as

> the biggest planning application ever submitted in Britain. The developer, Peel Holdings, plans to regenerate 60 hectares of the waterfront, in close proximity to the Three Graces, with gigantic skyscrapers (e.g. a 55-storey Shanghai Tower) and 1.7 million square metres of offices, homes and shopping.

The proposed Shanghai Tower, named in honour of Liverpool's aspirational relationship with the Chinese city, is part of an "aggressive entrepreneurial policy" (Attademo, 2013, p163) designed to encourage financial investment from China. If the project goes ahead, the city's cherished UNESCO world heritage status will be put at risk (Boland, 2013), as "the Pier Head stays on the right of the picture, just like a tiny memory of an irrelevant past" (Attademo, 2013, p165).

Whilst some commentators can see the potential benefits of the project to the city and its population, in the form of jobs, new dwellings, and restoration of a "general eyesore" (Taylor, 2013) in the form of brownfield land, others have raised concerns. Boland (2013, p261) highlights the potential for further (social) control of quasi-public/quasi-private space, and notes that "in recent years certain forms of cultural and political expression became illegal in the city centre due to fears of crime, risk and harm" – the Liverpool Waters project could well prove to be the catalyst for a spatial extension of such proscription.

Conclusion

> Everything is awesome
> Everything is cool when you're part of a team
> Everything is awesome
> When you're living out a dream.
>
> (The Lego Movie, 2014)

In many regards Liverpool has changed considerably since its heritage was 'discovered' in the early 1980s – derelict land has been reclaimed and restored, beautiful buildings are cherished instead of ignored, and The Beatles are located against a backdrop of many thousands of bright, funny, creative Scousers rather than as one-offs. Much is made of the distinctive nature of the city, in that there is no other place quite like it; its heritage is portrayed as unique, just like its populace. However, Liverpool remains in many ways just another city; post-industrial, post-economic freefall, and post-regeneration – a regeneration which has consolidated its similarities to other cities rather than highlighted its differences, bringing with it all the hallmarks of a European city worth visiting, from open top buses and walking tours to a Starbucks/Café Nero/Pret a Manger on, seemingly, every corner. In this sense, Liverpool feels like the set of *The Lego Movie* – a place constructed using the same materials and the same instructions as many other cities, and a place where inhabitants are also provided with instructions for living: "To fit in, have everyone like you and always be happy" (ibid.). The Lego-land instructions for living are reflective of the habitus that provides the 'rules' for the people of Liverpool, and the ideology that the contemporary Liverpool is indeed 'awesome' (as referenced in the quote at the top of this section); the same habitus that has led my participants to connect their understanding of home, in the context of contemporary regeneration initiatives, to the city centre and waterfront rather than their immediate district or neighbourhood. Chapter Six begins to explore the participants' relationship with their home city, and how it has shifted over time in accordance with prevailing development and regeneration climates.

Notes

1 The Chief Constable of Merseyside Police during these events, Kenneth Oxford, was clear in his view that "it is a matter of historical fact that Liverpool has been beset by problems of violence and disorder throughout the centuries" (cited by Coleman, 2004, p104).

2 Confidential documents of national importance or containing sensitive material are usually only registered with National Archives in the UK after 30 years have passed (National Archives, n.d.).

3 Frost and North (2013, p19) cite Mike Hennessey, who was employed at the time in Renshaw Street Unemployment Benefit Office: "Monday morning after a close down you'd be inundated with new people making claims. I remember people coming in who had never signed on before – you could see them totally lost and bewildered…people's faces coming in, the despair and [expression] of 'what the hell happens next?'"

4 Bleasdale's *Boys from the Blackstuff* was a television drama series which was first aired by the BBC in 1982 and was repeated in full with unprecedented speed in 1983 (Millington and Nelson, 1986). Each of the five episodes focused on a different male character and their struggle with long-term unemployment, and the drama is widely regarded as a "state of the nation" piece which disclosed the reality of post-industrial poverty in both a sympathetic and overtly political manner (Hallam, 2007, p184).
5 *Up the Junction* is a 1965 BBC 'Wednesday Play' based on a collection of short stories about 'slum' life in London written by Nell Dunn. The play was considered groundbreaking; first, for the range of social issues it tackled (domestic violence, illegal abortion, and criminality); second, for its use of location rather than studio filming to better reflect the realism of the material (MacMurragh-Kavanagh and Lacey, 1999).
6 *Cathy Come Home*, was another 'Wednesday Play', first aired in 1966 and is widely accepted as an early example of a 'campaigning' piece of television which sought to highlight and, crucially, ameliorate the contemporary problem of homelessness (Paget, 1999).
7 Couch (2003, p106) dates the emergence of drives towards tourism back to this period: "The potential of the city to be recognised with a modest policy to retain and enhance existing tourism assets and to protect the tourist potential of areas. In addition new attractions were to be encouraged, especially those on the Mersey waterfront".
8 Taylor (2009, p23) cites Terry Davenport, former director of BDP, the landscape architecture group who oversaw the Liverpool One project: "I still recall watching a classic piece of Granada TV coverage…when local architect Ken Martin…argued the case for the retention of the Albert Dock. At the time, and quite remarkably, developers proposed the wholesale demolition of the existing buildings. Ken argued his case around a beautifully set out table with 12-piece dinner service and then, towards the end of his statement, proceeded to walk around it, smashing all of the plates with a hammer! This was fairly compelling viewing and struck home, I'm sure for all who watched, the importance of protecting our heritage from the irretrievable damage of the bulldozer".
9 Frost and North (2013, p126) interviewed former Labour MP for Liverpool Walton Peter Kilfoyle, who described the phenomenon of 'workerism' as stemming from the desire to "look after one's own", i.e., to ensure as far as possible that family members, neighbours, even people of the same religion be prioritised before others. Whilst this "kith and kin recruitment" (Meegan, 1989, p215) is perhaps understandable within a climate of limited work and casualised labour, it is also exclusionary towards 'outsiders' (ibid.).
10 Derek Hatton, former deputy leader of Liverpool City Council, was a key figure and spokesperson for the Militant movement in Liverpool (Parkinson, 1985).
11 Blandy and Hunter (2012) note that the 'right to buy' council houses for council tenants established under the Housing Act 1980 is very much in keeping with the expanding neoliberal project of the day. On this basis, the common use of language here is interesting; 'right to buy' suggests that it is *right* i.e., correct, appropriate, proper to buy one's home rather than rent it.
12 The Heysel Stadium Disaster occurred on 29 May 1985 when 39 football fans were killed before and during the European Cup final between Liverpool FC and Juventus FC. Whilst the disrepair of the stadium and the 'mixed' seating arrangements within it were contributory factors, the blame for the disaster was, in the main, attributed to the 'hooliganism' of the "beasts from Liverpool" (Chisari, 2004, p203).
13 The Hillsborough Disaster of 15 April 1989 brought about the deaths of 96 Liverpool fans through asphyxiation as a result of crushing within the central pen of the Leppings Lane end of Hillsborough Stadium, Sheffield. For many years official blame for these events was attributed to Liverpool football fans via several mechanisms (Scraton, 2009); it was not until 26 April 2016 that the fans and victims were exonerated by an inquest jury which found conclusively that the tragedy was caused directly by police actions and mismanagement (Conn, 2016).

14 Objective 1 funding can be awarded by the European Union to areas where gross domestic product per head is less than 75% of the European Union average (Jones and Skilton, 2014).
15 Merseyside's initial funding period ran from 1993 to 1999; the area was granted a second period of funding from 2000 to 2006 (Gripaios and Bishop, 2006).
16 This phrase is used to describe buildings and structures that appear to be 'plonked' down in a random place without thought, planning, or consideration (Rodwell, 2014).
17 Several streets in Liverpool city centre started life as 'rope walks' – long, narrow passages which facilitated the manufacture of long ropes used on the docks and at sea. Liverpool's Rope Walks Project seeks to conserve the physical nature of these streets to reflect their history (Rodwell, 2015).
18 Although only redeveloped in the 1990s, the Paradise Street bus station and car park was demolished in 2005 in preparation for the building of the Liverpool One Development (Barratt, 2005).
19 'Quality of life' and 'zero tolerance' policing strategies purport to serve the best interests of the law-abiding majority by both improving the look, feel, and ambience of a given area by aggressively policing such 'anti-social behaviour' as graffiti, unauthorised loud music, begging, the public drinking of alcohol, and general non-approved street activity (Innes, 1999).
20 In the course of wide reading for this project, I have come across the phrases 'world class city' and 'world class status' in relation to Liverpool many times, yet I have not been able to find any real discussion on/explanation of these phrases and what they actually *mean* in any measurable sense. It seems fair to conclude that they are an example of meaningless jargon associated with statecraft and regeneration processes.
21 Long-term unemployment is often referred to, particularly in policy literature, as 'worklessness' – an approach which "suggest[s] that people are unemployed because of their values, attitudes and behaviours rather than because of a shortage of jobs. In simple terms, they imply people prefer a life on welfare benefits to working for a living" (Shildrick et al., 2012, p5). This choice of terminology, used frequently for example in the 2011 report by Heseltine and Leahy, indicates a particular political standpoint in relation to long-term unemployment and the social marginality it can lead to.
22 According to the Institute for Sustainable Communities, "a sustainable community is one that is economically, environmentally, and socially healthy and resilient. It meets challenges through integrated solutions rather than through fragmented approaches that meet one of those goals at the expense of the others. And it takes a long term perspective – one that's focused on both the present and the future, well beyond the next budget or election cycle".
23 The New Deal for Communities programme was instigated under the Blair government in 1998 (Lawless, 2006). The Kensington New Deal, focused on the Kensington district in central Liverpool, initially promised to provide for the demolition of 900 houses, the building of 500 new ones, a new health centre, improvements to the look and feel of the area (e.g., improved street lighting) and new retail developments (Atkinson and Maliene, 2015).
24 Peel Holdings, chaired by the reclusive John Whittaker and registered in the tax haven of the Isle of Man, would go on to become the single most influential private organisation in the North West of England; owning vast swathes of land and various developments, several of which are collaborative endeavours with the public sector (Harper, 2013).
25 According to Traynor (2009), Sir Bob Scott, who led the bid, later admitted to manipulating the bid judges by 'planting' people who they thought would make a good impression in the pubs that the judges visited: "We worked out who they were and we bloody made sure who the people in the pub were when they arrived, and that made quite an impact…we did not take this group of people to rooms full of men in suits, we took

them to schools, got them down on the floor with kids. We really got the impression across in every visual way we could that there was real public support".

26 In 2015 the University of Warwick published a report entitled 'Enriching Britain: Culture, creativity and growth'. The document, intended by its authors to be a "blueprint for Britain's cultural and creative enrichment", makes no attempt at all to provide a frame of reference and presents a one-dimensional perspective regarding what could, or should, be viewed as culture; it is presented as a wholly 'taken for granted' concept. There is no mention of such 'low' cultural activities as, for example, darts, Northern Soul, karaoke, holy communion celebrations, or other activities that are both commonplace and popular in Liverpool and other cities.

27 One particularly egregious example is the diversion of Neighbourhood Renewal Fund money, intended for community projects in Norris Green, Everton, and Toxteth, to fund the Capital of Culture shop, based in Whitechapel (city centre), which ultimately made a financial loss (Furmedge, 2008).

28 'Superlambanana', a sculpture designed by Taro Chiezo, has become iconic of Liverpool since it was commissioned in 1998 (Hannam and Knox, 2010).

29 Lee (2007, p152) defines the term thus: "The *flaneur* is the man who wanders the city, the man who finds pleasure in being away from home, freedom in being non-locatable". In this sense Boland "walks the streets of the city in order to understand it, so much so that he ultimately represents and even embodies the public nature of the city" (Kinsella, 2012, p123).

30 There is an inherent assumption in cultural studies that unthinkingly equates 'high' and 'low' culture with different social classes; for example, Rai and Panna (2015, p6) make this sweeping statement: "The different economic and social strata of people in a society select and prefer to access different convenient forms of mass media. The lower income groups prefer audio-visual form of media to the print media".

31 Streets and squares in the Abercromby area, for example, Faulkner Square, have been 'heritagised' with the edition of faux-Georgian street lamps, to enhance the 'authenticity' of the area (Speake and Fox, 2006).

32 Lovering (2007, p344) deftly and disdainfully demonstrates the often flimsy link between heritage, culture, and regeneration, for example: "A new shopping mall in Cardiff will provide a 'Welsh experience' (i.e., it features pictures of Catherine Zeta Jones)".

33 Liv and Liz were developed as characters for Liverpool One by PR and advertising firm McCann Manchester, whose brief was to make Liverpool One stand out against other shopping destinations: "The TV campaign features two eccentric characters, Liv and Liz, who are Liverpool's very own iconic Liver Birds – the famous statues perched on top of the Royal Liver Building. These symbols of the city are brought to life through animation, transforming them into two stylish, fun loving, straight talking girls".

Bibliography

Allen C (2008) *Housing Market Renewal and Social Class Formation.* Routledge: Abingdon. doi:10.4324/9780203932742.

Allt N (2008a) 'The un-English disease' in Allt N (ed) *The Culture of Capital.* Liverpool University Press: Liverpool, pp 13–42.

Allt N (2008b) 'Liverpool DNA' in Allt N (ed) *The Culture of Capital.* Liverpool University Press: Liverpool, pp 43–48.

Allt N (2008c) 'Sintroduction' in Allt N (ed) *The Culture of Capital.* Liverpool University Press: Liverpool, pp 1–8.

Anderson J (2015) 'Liverpool on a world stage' *Conference address Liverpool International Waterfront Forum*, 3–4 June 2015, Liverpool.

Atkinson I and Maliene V (2015) 'Challenges of English planning policies: Regeneration and sustainable communities' in Hepperle E, Dixon-Gough R, Mansberger R, Paulsson J, Reuter F and Yilmaz M (eds) *Challenges for Governance Structures in Urban and Regional Development*. vdf Hochschulverlag: Zurich, pp 235–249.

Attademo A (2013) 'Liverpool@Shanghai: The waterfront as a *brandscape* in Liverpool Waters case study' *Sea and the City* 11 (2) pp 157–170. doi:10.6092/2281-4574/2046.

Aughton P (2008) *Liverpool: A People's History*, 3rd ed. Carnegie Publishing: Lancaster.

Avery P (2007) 'Born again: From dock cities to cities of culture' in Smith M (ed) *Tourism, Culture and Regeneration*. CAB International: Wallingford, pp 151–162. doi:10.1079/9781845931308.0000.

Bartoletti R (2010) '"Memory tourism" and commodification of nostalgia' in Burns P, Palmer C and Lester J (eds) *Tourism and Visual Culture, Volume 1: Theories and Concepts*. CABI Publishing: Wallingford, pp 23–42. doi:10.1079/9781845936099.0023.

Belchem J and MacRaild D (2006) 'Cosmopolitan Liverpool' in Belchem J (ed) *Liverpool 800: Culture, Character and History*. Liverpool University Press: Liverpool, pp 311–391.

Balderstone L, Milne G and Mulhearn R (2014) 'Memory and place on the Liverpool waterfront in the mid-twentieth century' *Urban History* 41 (3) pp 478–496. doi:10.1017/S0963926813000734.

Barratt T (2005) 'Halfway to paradise?' *Liverpool Echo*, 28 November 2005. Available at http://www.liverpoolecho.co.uk/news/liverpool-news/halfway-to-paradise-3525809. Accessed 18 July 2016.

Bauman Z (2011) *Culture in a Liquid Modern World*. Polity: Cambridge.

Bayley S (2010) *Liverpool: Shaping the City*. RIBA Press: London.

Bennett J (2007) 'Manners, mores and musicality: An interview with Willy Russell' in Jones D and Murphy M (eds) *Writing Liverpool: Essays and Interviews*. Liverpool University Press: Liverpool, pp 228–238. doi:10.5949/upo9781846314476.017.

Blandy S and Hunter C (2012) 'The right to buy: Examination of an exercise in allocating, shifting and re-branding risks' *Critical Social Policy* 33 (1) pp 17–36. doi:10.1177%2F0261018312457869.

Boland P (1996) 'Institutional mechanisms and regional development in Merseyside: Objective 1 status' in Alden J and Boland P (eds) *Regional Development Strategies: A European Perspective*. Jessica Kingsley: Abingdon, pp 107–128. doi:10.4324/9781315000220.

Boland P (1999) 'Merseyside and Objective 1 status, 1994–1999: Implications for the next programming period' *Regional Studies* 33 (8) pp 788–792.

Boland P (2000) 'Urban governance and economic development: A critique of Merseyside and Objective 1 status' *European Urban and Regional Studies* 7 (3) pp 211–222.

Boland P (2007) 'Unpacking the theory – Policy interface of local economic development: An analysis of Cardiff and Liverpool' *Urban Studies* 44 (5/6) pp 1019–1039. doi:10.1080/00420980701320736.

Boland P (2008) 'The construction of images of people and place: Labelling Liverpool and stereotyping Scousers' *Cities* 25 pp 355–369. doi:10.1016/j.cities2008.09.003.

Boland P (2010a) '"Capital of Culture – you must be having a laugh!" Challenging the official rhetoric of Liverpool as the 2008 European cultural capital' *Social and Cultural Geography* pp 627–645. doi:10.1080/14649365.2010.508562.

Boland P (2010b) 'Sonic geography, place and race in the formation of local identity: Liverpool and Scousers' *Geografiska Annaler Series B Human Geography* 92 (1) pp 1–22. doi:10.1111/j.1468-0467.2010.00330.x.

Boland P (2013) 'Sexing up the city in the international beauty contest: The performative nature of spatial planning and the fictive spectacle of place branding' *Town Planning Review* 84 (2) pp 251–274. doi:10.2307/23474330.

Browne A (2014) 'Have you joined the love locks craze? Share your pictures with us' *Liverpool Echo*, 7 October 2014. Available at http://www.liverpoolecho.co.uk/whats-on/whats-on-news/you-joined-love-locks-craze-7897599. Accessed 6 June 2017.

Burnett J (1986) *A Social History of Housing 1815–1985*, 2nd ed. Methuen: London.

Butler D, Adonis A and Travers T (1994) *Failure in British Government: The Politics of the Poll Tax*. Oxford University Press: Oxford.

Chisari F (2004) '"The cursed cup": Italian responses to the 1985 Heysel disaster' *Soccer & Society* 5 (2) pp 201–218. doi:10.1080/1466097042000235218.

Coleman R (2004) *Reclaiming the Streets: Surveillance, Social Control and the City*. Willan: Cullompton.

Coleman R and Hancock L (2009) 'Culture and curfew in Fantasy City: Whose time, whose place?' *Nerve* 14 pp 12–13.

Coming Home Liverpool (2017) *Empty Homes*. Available at https://cominghomeliverpool.wordpress.com/category/empty-homes/. Accessed 2 June 2017.

Conn D (2016) 'Hillsborough inquests jury rules 96 victims were unlawfully killed' *The Guardian*, 26 April 2016. Available at https://www.theguardian.com/uk-news/2016/apr/26/hillsborough-inquests-jury-says-96-victims-were-unlawfully-killed. Accessed 22 June 2016.

Connolly M (2013) 'The 'Liverpool Model(s)': Cultural planning, Liverpool and Capital of Culture 2008' *International Journal of Cultural Policy* 19 (2) pp 162–181. doi:10.1080/10286632.2011.638982.

Couch C (2003) *City of Change and Challenge: Urban Planning and Regeneration in Liverpool*. Ashgate: Aldershot. doi:10.4324/9781315197265.

Courtman S (2007) 'Culture is ordinary: The legacy of the Scottie Road and Liverpool 8 writers' in Jones D and Murphy M (eds) *Writing Liverpool: Essays and Interviews*. Liverpool University Press: Liverpool, pp 194–209. doi:10.5949/upo9781846314476.015.

Cousins S (2013) 'Broad Lane: From estate hell to housing heaven' *CM:Magazine of the Chartered Institute of Building*, 2 June 2003. Available at http://www.constructionmanagermagazine.com/features/estate-hell-housing-heaven/. Accessed 29 September 2016.

Cox T and O'Brien D (2012) 'The "Scouse wedding" and other myths: reflections on the evolution of a "Liverpool model" for urban regeneration' *Cultural Trends* 21 (2) pp 93–101 doi:10.1080/09548963.2012.674749.

Culture Liverpool (2017) *67–17: 50 Summers of Love*. Available at https://www.cultureliverpool.co.uk/summer-of-love/. Accessed 8 May 2017.

Czarniawska B (2002) *A Tale of Three Cities: Or the Glocalization of City Management*. OUP: Oxford. doi:10.1093/acprof:oso/9780199252718.001.0001.

Davis H and Bourhill M (1997) '"Crisis": The demonization of children and young people' in Scraton P (ed) *'Childhood' in 'Crisis'?* UCL: London, pp 28–57.

Devaney C (2011) 'Scouse and the city: Radicalism and identity in contemporary Liverpool' in Belchem J and Biggs B (eds) *Liverpool: City of Radicals*. Liverpool University Press: Liverpool, pp 172–183.

Dow C (1998) *Major Recessions: Britain and the World, 1925–1995*. OUP: Oxford.

Du Noyer P (2002) *Liverpool: Wondrous Place: Music from the Cavern to the Coral*. Virgin Books: London.

Erll A (2011) *Memory in Culture*. Palgrave: Basingstoke.
Evans K (2003) 'The contemporary city: A critical perspective' in Munck R (ed) *Reinventing the City: Liverpool in Comparative Perspective*. Liverpool University Press: Liverpool, pp 23–35. doi:10.5949/liverpool/9780853237976.003.0002.
Evans R (2002) 'The Merseyside Objective One programme: Exemplar of coherent city-regional planning and governance or cautionary tale?' *European Planning Studies* 10 (4) pp 495–517. doi:10.1080/09654310220130202.
Fagan G (2004) *Liverpool: In a City Living*. Countyvise: Birkenhead.
Foley M, McGillivray D and McPherson G (2012) *Event Policy: From Theory to Strategy*. Routledge: Abingdon. doi:10.4324/9780203806425.
Frost D and North P (2013) *Militant Liverpool: A City on the Edge*. Liverpool University Press: Liverpool. doi:10.5949/liverpool/9781846318634.001.0001.
Frost D and Phillips R (2011) (eds) *Liverpool '81: Remembering the Riots*. Liverpool University Press: Liverpool.
Furmedge P (2008) 'The regeneration professionals' in Allt N (ed) *The Culture of Capital*. Liverpool University Press: Liverpool.
Garcia B, Melville R and Cox T (2010) *Creating an Impact: Liverpool's Experience as European Capital of Culture*. Impacts 08: Liverpool.
Getting By? (2015) *Getting By? A Year in the Life of 30 Working Families in Liverpool*. Available at www.gettingby.org.uk. Accessed 11 May 2015.
Graham P and Clarke J (2001) 'Dangerous places: Crime and the city' in Muncie J and McLaughlin E (eds) *The Problem of Crime*, 2nd ed. Sage: London, pp 152–190.
Gripaios P and Bishop P (2006) 'Objective one funding in the UK: A critical assessment' *Regional Studies* 40 (8) pp 937–951. doi:10.1080/00343400600877839.
Hallam J (2007) '"We are a city that just likes to talk": An interview with Alan Bleasdale' in Jones D and Murphy M (eds) *Writing Liverpool: Essays and Interviews*. Liverpool University Press: Liverpool, pp 184–193. doi:10.5949/UPO9781846314476.014.
Hannam K and Knox D (2010) *Understanding Tourism: A Critical Introduction*. Sage: London. doi:10.4135/9781446288528.
Harrison J (2014) 'Rethinking city-regionalism as the production of new non-state spatial strategies: The case of Peel Holdings Atlantic Gateway Strategy' *Urban Studies* 51 (11) pp 2315–2335. doi:10.1177%2F0042098013493481.
Harper T (2013) 'The biggest company you've never hear of: Lifting the lid on Peel Group – The property firm owned by reclusive tax exile John Whittaker' *The Independent*, 18 October 2013. Available at http://www.independent.co.uk/news/uk/home-news/the-biggest-company-youve-never-heard-of-lifting-the-lid-on-peel-group-the-property-firm-owned-by-8890201.html. Accessed 29 September 2016.
Herbert I (2006) 'Liverpool's city of culture plans in tatters as boss quits' *The Independent*, 4 July 2006. Available at http://www.independent.co.uk/news/uk/this-britain/liverpools-city-of-culture-plans-in-tatters-as-boss-quits-6096259.html-0. Accessed 27 May 2017.
Heseltine M and Leahy T (2011) *Rebalancing Britain: Policy or Slogan? Liverpool City Region – Building on its Strengths: An Independent Report*. Department for Business, Education and Skills: London.
Higginson S (2008) 'Outsourced Liverpool' in Allt N (ed) *The Culture of Capital*. Liverpool University Press: Liverpool, pp 51–56.
Hobson D (2008) *Channel 4: The Early Years and the Jeremy Isaacs Legacy*. IB Tauris: London.
Holmes C (2003) *Housing, Equity and Public Policy*. Institute of Public Policy Research: London.

Ilan J (2015) *Understanding Street Culture: Poverty, Crime, Youth and Cool*. Palgrave: London.

Innes M (1999) '"An iron fist in an iron glove": The zero tolerance policing debate' *The Howard Journal of Criminal Justice* 38 (4) pp 397–410. doi:10.1111/1468-2311.00143.

Institute for Sustainable Communities (nd) *Definition of Sustainable Community*. Available at http://www.iscvt.org/impact/definition-sustainable-community/. Accessed 20 July 2016.

Ironside M and Seifert R (2000) *Facing Up to Thatcherism: The History of NALGO* 1979–1993. Oxford University Press: Oxford. doi:10.1093/acprof:oso/9780199240753.001.0001.

Irwin M (2016) '*The Rag Trade:* "Everybody out!" Gender, politics and class on the factory floor' in Kamm J and Neumann B (eds) *British TV Comedies: Cultural Concepts, Contexts and Controversies*. Palgrave Macmillan: Basingstoke, pp 66–79. doi:10.1057/9781137552952_5.

Jones C (2008) 'Liverpool's art lovers ready to go superlambananas' *Liverpool Echo*, 14 June 2008. Available at http://www.liverpoolecho.co.uk/news/liverpool-news/liverpools-art-lovers-ready-go-3483661. Accessed 27 May 2017.

Jones M and Skilton L (2014) 'An analysis of labour market outcomes in the European Union Objective One funding area in Great Britain' *Regional Studies* 48 (7) pp 1194–1211. doi:10.1080/00343404.2012.697992.

Jones P (2016) 'Architecture as spatio-temporal fix: The case of Liverpool One' Paper presented to *Urban Studies Liverpool Symposium*, Blackburne House, Liverpool, 16 September 2016.

Kemp R (2015) 'Keynote speech' *Liverpool International Waterfront Forum*, 3–4 June 2015, Liverpool.

Kensington Regeneration (nd) *Kensington Regeneration Hailed as a Success*. Available at http://www.kensingtonregeneration.org/news_and_events/press/kensington_regeneration_hailed_as_a_success/index.html. Accessed 9th April 2017.

Kensington Regeneration (2008) *Deliverington: Kensington Regeneration Delivery Plan 2008/9*. Available at http://kensingtonregeneration.org/images/uploads/delivery_plan.pdf. Accessed 2 June 2017.

Kinsella C (2011) 'Welfare, exclusion and rough sleeping in Liverpool' *International Journal of Sociology and Social Policy* 31 (5/6) pp 240–252. doi:10.1108/01443331111141246.

Kinsella C (2012) 'Re-locating Fear on the Streets: Homelessness, victimisation and fear of crime' *European Journal of Homelessness* 6 (2) pp 121–136.

Lavergna A, Lombardo R and Sergi A (2013) 'Organised crime in 3 regions: Comparing the Veneto, Liverpool and Chicago' *Trends in Organised Crime* 16 (3) pp 265–285. doi:10.1007/s12117-013-9189-1.

Lawless P (1989) *Britain's Inner Cities*, 2nd ed. Paul Chapman Publishing: London.

Lawless P (2006) 'Area-based urban interventions: Rationale and outcomes: The New Deal for Communities programme in England' *Urban Studies* 43 (11) pp 1991–2011. doi:10.1080%2F00420980600897859.

Lee M (2007) *Inventing fear of crime: Criminology and the politics of anxiety*. Willan: Cullompton

Levitas R (1998) *The Inclusive Society? New Labour and Social Exclusion*. Macmillan: Basingstoke.

Levitas R (2006) 'The concept and measurement of social exclusion' in Pantazis C, Gordon D and Levitas R (eds) *Social Exclusion in Britain*. The Policy Press: Bristol, pp 123–160.

Liu Y (2015) 'Event-led strategy for cultural tourism development: The case of Liverpool as the 2008 European Capital of Culture' *The Planning Review* 51 (2) pp 28–40. doi:10.1080/02513625.2015.1064645.

Lovering J (2007) 'The relationship between urban regeneration and neoliberalism: Two presumptuous theories and a research agenda' *International Planning Studies* 12 (4) pp 343–366. doi:10.1080/13563470701745504.

Low S (2017) *Spatialising Culture: The Ethnography of Space and Place*. Routledge: Abingdon. doi:10.4324/9781315671277.

MacMurragh-Kavanagh M and Lacey S (1999) 'Who framed theatre? The "moment of change" in British TV drama' *New Theatre Quarterly* 15 (1) pp 58–74. doi:10.1017/S0266464X00012653.

McCann Manchester (nd) *Liverpool One: Creating Standout for a Shopping and Leisure Destination*. Available at http://www.mccannmanchester.com/work/case-studies/liverpool-one/. Accessed 6 June 2017.

McColgan C (2015) 'Place making on the waterfront' Unpublished conference paper *Liverpool International Waterfront Forum*, 3–4 June 2015, Liverpool.

Mah A (2014) *Port Cities and Global Legacies: Urban Identity, Waterfront Work and Radicalism*. Palgrave MacMillan: Basingstoke.

Mahler S (2013) *Culture as Comfort*. Pearson: New Jersey.

Marren B (2016) *We Shall Not Be Moved: How Liverpool's Working Class Fought Redundancies, Closures and Cuts in the Age of Thatcher*. Manchester University Press: Manchester. doi:10.7228/manchester/9780719095764.001.0001.

Meegan R (1989) 'Paradise postponed: The growth and decline of Merseyside's outer estates' in Cooke P (ed) *Localities: The Changing Face of Urban Britain*. Unwin Hyman: London, pp 198–234.

Meegan R (1990) 'Merseyside in crisis and conflict' in Harlow E, Pickavance C and Urry J (eds) *Place, Policy and Politics: Do Localities Matter?* Unwin Hyman: London, pp 87–107.

Meegan R (1994) 'A "Europe for the regions?" A view from Liverpool on the Atlantic periphery' *European Planning Studies* 2 pp 59–80.

Meegan R (2003) 'Urban regeneration, politics and social cohesion: The Liverpool case' in Munck R (ed) *Reinventing the City: Liverpool in Comparative Perspective*. Liverpool University Press: Liverpool, pp 53–79. doi:10.5949/liverpool/9780853237976.003.0004.

Meegan R and Mitchell A (2001) '"It's not community round here, it's neighbourhood": Neighbourhood change and cohesion in urban regeneration policies' *Urban Studies* 38 (12) pp 2167–2194. doi:10.1080%2F00420980120087117.

Millington B and Nelson R (1986) *'Boys from the Blackstuff': The Making of TV Drama*. Comedia: London.

Minton A (2012) *Ground Control: Fear and Happiness in the 21st Century*. Penguin: London.

Moore K (1997) *Museums and Popular Culture*. Cassel: London.

Morgan N (2014) *The Heroin Epidemic of the 1980s and 1990s and its Effect on Crime Trends – Then and Now*. Home Office Research Report 79. Available at https://www.gov.uk/government/uploads/system/uploads/attachment_data/file/332952/horr79.pdf. Accessed 28 June 2016.

Muchnick D (1970) *Urban Renewal in Liverpool*. The Social Administration Research Trust: Birkenhead.

Mulhearn D (2003) 'Given the boot' *The Guardian*, 28 May 2003. Available at https://www.theguardian.com/society/2003/may/28/regeneration.communities. Accessed 29 September 2016.

Murden J (2006) 'City of change and challenge: Liverpool since 1945' in Belchem J (ed) *Liverpool 800: Culture, Character and History*. Liverpool University Press: Liverpool, pp 393–485.

National Archives (nd) *The Cabinet Papers 1915–1988* [Online]. Available at http://www.nationalarchives.gov.uk/cabinetpapers/cabinet-gov/meetings-papers.htm. Accessed 11 June 2016.

Paget D (1999) '"Cathy Come Home" and "accuracy" in British television drama' *New Theatre Quarterly* 15 (1) pp 75–90. doi:10.1017/S0266464X00012665.

Parkinson M (1985) *Liverpool on the Brink: One City's Struggle against Government Cuts*. Policy Journals: Hermitage.

Parry J (2016) 'Liverpool in housing crisis – With only one home built for every eight new arrivals' *Liverpool Echo*, 10 September 2016. Available at http://www.liverpoolecho.co.uk/news/liverpool-news/liverpool-housing-crisis-only-one-11866028. Accessed 2 June 2017.

Peel D, Lloyd G and Lord A (2007) 'Business improvement districts and the discourse of contractualism' *European Planning Studies* 17 (3) pp 401–422. doi:10.1080/09654310802618044.

Peel Property & Land (2015) *Visionary, Sustainable, Dynamic: Delivering Sustainable Regeneration*. Available at http://peelgroup.global/investments/transport-logistics/aviation. Accessed 29 September 2016.

Potschulat M (2016) 'Constructing "the student experience" through housing: A case study of Liverpool' Paper presented to *Urban studies Liverpool Symposium*, Blackburne House, Liverpool, 16 September 2016.

Power A and Wilson W (2000) *Social Exclusion and the Future of Cities*. Centre for Analysis of Social Exclusion: London. Available at http://papers.ssrn.com/sol3/papers.cfm?abstract_id=1158926. Accessed 18 July 2016.

Rai R and Penna K (2015) *Introduction to Culture Studies*. Himalaya Publishing: Mumbai.

Roberts L (2010) 'World in one city: Surrealist geography and time-space compression in Alex Cox's Liverpool' *Tourism and Visual Culture* 1 pp 200–215. doi:10.1079/9781845936099.0200.

Roberts L (2012) *Film, Mobility and Urban Space: A Cinematic Geography of Liverpool*. Liverpool University Press: Liverpool. doi:10.5949/upo9781846317248.

Robson B (1988) *Those Inner Cities: Reconciling the Social and Economic Aims of Urban Policy*. Clarendon Press: Oxford.

Rodwell D (2014) 'Negative impacts of World heritage branding: Liverpool – An unfolding tragedy?' *Primitive Tider* 14 pp 19–35. doi:10.1179/1756750515Z.00000000066.

Rodwell D (2015) 'Liverpool: Heritage and development – Bridging the gap?' in Oevermann H and Mieg H (eds) *Industrial Heritage Sites in Transformation: Clash of Discourses*. Routledge: Abingdon, pp 29–46.

Scraton P (2009) *Hillsborough: The Truth*, 20th anniversary ed. Mainstream Publishing: Edinburgh.

Shaw S (2010) *Parents, Children, Young People and the State*. Open University Press: Maidenhead.

Shildrick T, MacDonald R, Furlong A, Roden A and Crow R (2012) *Are 'Cultures of Worklessness' Passed Down the Generations?* Joseph Rowntree Foundation. Available at http://wbg.org.uk/pdfs/worklessness-families-employment-full.pdf. Accessed 20 July 2016.

Speake J and Fox V (2006) *Discovering Cities: Liverpool*. Geographical Association: Sheffield.

Tallon A (2013) *Urban Regeneration in the UK*, 2nd ed. Routledge: Abingdon. doi:10.4324/9780203802847.

Taylor D (2009) *Liverpool: Regeneration of a City Centre*. BDP: Manchester.

Taylor K (2011) 'From the ground up: Radical Liverpool now' in Belchem J and Biggs B (eds) *Liverpool: City of Radicals*. Liverpool University Press: Liverpool, pp 159–171.

Taylor K (2013) 'The breathtaking potential of Liverpool (and Wirral) Waters' *The Guardian*, 7 March 2013. Available at https://www.theguardian.com/uk/the-northerner/2013/mar/07/liverpool-architecture-peelgroup-mersey-wirral-docks-regeneration. Accessed 15 April 2016.

The Lego Movie [DVD] (2014) Chris Miller Phil Lord dirs. USA: Village Roadshow Pictures, Animal Logic, Warner Bros Animation.

The Warwick Commission (2015) *Enriching Britain: Culture, Creativity and Growth*. University of Warwick: Coventry.

Timothy D (2011) *Cultural Heritage and Tourism: An Introduction*. Channel View Publications: Bristol.

Travis A (2011) 'Thatcher government toyed with evacuating Liverpool after 1981 riots' *The Guardian*, 30 December 2011. Available at https://www.theguardian.com/uk/2011/dec/30/thatcher-government-liverpool-riots-1981. Accessed 11 June 2016.

Traynor L (2009) 'Bob Scott: How I "tricked" 08 culture [sic] of culture judges' *Liverpool Echo*, 20 January 2009. Available at http://www.liverpoolecho.co.uk/news/liverpool-news/bob-scott-how-tricked-08-3462303. Accessed 5 June 2017.

Tulloch A (2011) *The Story of Liverpool*. The History Press: Stroud.

Waddington M (2013) 'Liverpool Militant 30 years on: Roy Gladden and the rebellion within the rebellion' *The Liverpool Echo*, 10 May 2013. Available at http://www.liverpoolecho.co.uk/news/liverpool-news/liverpool-militant-30-years-on-3572383. Accessed 21 June 2016.

Wright J (2015) 'The Mathew Street Festival – 40 pictures from a 21 year party' *The Liverpool Echo*, 20 August 2015. Available at http://www.liverpoolecho.co.uk/whats-on/mathew-street-festival-40-pictures-9897472. Accessed 7 July 2016.

6 The Liverpool home in the 'forward-facing' era

This chapter explores the experiences of home in Liverpool up until the late 1970s. Much of what is discussed here pertains to experiences and memories from the 1950s onwards, although some testimony recalls earlier times, for example during the Second World War. Findings here have been grouped into themes which emerged from the data and reflect the different stages of regeneration during this period, their impact, and how they were responded to and reflected on by those for whom Liverpool is home. 'Pen portraits' for each of the respondents can be found on page 211.

People loved the home that they had

With some limited exceptions,[1] the research participants spoke very warmly of their childhood homes, both in terms of the dwelling they were housed in and its immediate environs. Jill, for example, describes Pecksniff Street in Toxteth, where she lived until she was seven, as 'very much a, people say like Coronation Street, but it was in a way because it was a case of everyone knew everyone else, and yes it [was] a good sort of community spirit there'. Similarly, Irene recounts Minto Street in Kensington, her first home, to be 'very friendly, you could just go out and leave your door open, everybody looked after everybody, it was just a general community feeling, everybody knew everybody'. Lynne, who spent her early years in Seaton Street, Kensington, during the Second World War, insists that it was

> lovely, absolutely lovely, I know there were bombs dropping and everything else, but to children, no, it was a big adventure, people used to say 'oh God the kids', but the kids loved it, they were out collecting shrapnel, bits of bombs and things like that, and it was just lovely, and a real loving family.

Family is a central theme running through the memories of childhood homes of all the respondents. Several (Mary, Rita, Jill, Lynne, Christine, Peter, Enid, Pat, Maureen, Colette) had extended family living close by, including Teresa –

> Well I remember Pluto Street [Kirkdale] being lovely, it was just ordinary little houses ... but my nan lived in the same road, two aunties, and my nan's

> sister, my great aunt, lived next door. So it was very family oriented, all round the same [area].

Christine's first home in Lordens Road, Page Moss, backed on to her grandparents' house; her grandfather had made a gap in the dividing garden fence and fitted it with a gate, allowing immediate access between the homes. Roy spent his early years with several extended family members in one small house in Cotswold Street, Kensington,

> I absolutely loved it … there was my mum, dad, me, my nan, unfortunately my grandfather had died by the time I was born, and I think it was four of my aunties and uncles; Uncle Robert, Uncle Leonard, Maureen, trying to think who else now, no I think that was it, our Rita and Elsie had moved out already, so yeah there were three of them. So quite crowded, but fun.

Further, it appears common among the respondents to have operated as children within a localised social network, whereby neighbours were viewed as extended family and were referred to as 'auntie' or 'uncle'. Ronnie's testimony on his childhood in Maghull encapsulates this well:

> In the street [Hillary Crescent], I couldn't work out why I had so many aunties, the street was full of young women who didn't work, and collectively looked after us all. Sometimes your mum would call you in for your dinner, but it turned out you'd already had your dinner at Auntie Lorna's, and so it was very, I have very fond memories of that kind of communal living in those early days.

These quasi extended family networks facilitated financial help for those who needed it. Anne recalls the 'whip round' for families who had suffered a bereavement to cover burial and associated costs, while Maureen brings to mind her gratitude at acts of kindness, in the form of food or small amounts of money, when she was in financial hardship. Jill explains how her parents were able to take her and her older brother, Jack, who has Down's syndrome, to a specialist centre in Philadelphia, USA,[2] in the 1970s:

> I remember mum telling me that the neighbours, and everyone, friends, neighbours, actually raised money for them, they had a whip round for them to go. So again that speaks of a nice community spirit.

Respondents repeatedly demonstrated their feeling that everything they needed day to day could be found very close to home. Teresa describes Pluto Street as having everything close to hand:

> There was a pub on the end of the street, shop right opposite … there was a picture house called The Astoria, and there was a park just before, it's still there, but there's no swings and that, and we used to go there.

Similarly, Ronnie recalls the opportunity for freedom and adventure that his childhood home in Maghull afforded him: "Another memory of those days was how wide my radius was at a very early age, go anywhere as long as you were back in time for your tea, no one was bothered where you were". For Anne, the Marsh Lane (Bootle) area where she grew up was very much self-contained:

Anne: There were loads of shops there, it was nice, Marsh Lane, years ago.
John: The bottom end, all the shops were lit up of a night.
Clare: Alright, ok, so what types of shops were there? Was there everything?
Anne: Yeah, post office, chippy, pawn shop.
Clare: Clothes if you wanted clothes?
Anne: Yeah.
Clare: So probably then your mum didn't go much down Bootle?
Anne: No, she got everything she needed round Marsh Lane.

Both Rita and Irene remember Kensington having a range of impressive local amenities:

> There was a car showroom, there used to be a bingo down there as well, that's gone, where the Iceland and that is, there used to be an ice rink down there, I'm trying to think, what do you call the shops that sell fancy clothes? Boutiques, yeah, there was boutiques. (Rita)
>
> [Kensington had] every type of shop you could think of; we had three shoe shops, we had a Boots the Chemist, we had a Woolworths, we had Cooks the cake shop, we had a Cousins the cake shop … we had a shop called Hewson's which sold everything you could imagine; bedding, linen, crockery, cutlery … there was a milliner's, there was a ladies' dress shop, there was a children's dress shop. In 1972 it won the award for the best local shopping road! (Irene)

Thus, respondents mostly displayed contentment, pleasure and pride whilst remembering their early homes. What appeared to town planners in the mid-1960s as a dreary townscape (Couch, 2003), was viewed very differently by those who lived within it, and effectively provided respondents with a local ecology; with everything needed for everyday life close at hand. However, that is not to say that they did not recognise some of the problems they had, nor that nostalgia had caused them to lose sight of their structural position at the time and the hardships and difficulties these caused. Both Enid, in her house in Fishguard Street (Everton), and Pat, Maureen and Colette in their house in Wright Street (Scotland Road), experienced overcrowding as their big families overwhelmed their 'two up two down' dwellings; as Maureen recalled, 'we were always out, I don't remember being in the house very much when I lived in Scotland Road, because there was no room!' Ronnie's earliest home in Diana Street (Walton) was also crowded, as it was shared with another family.

Many of the respondents recalled their outside toilets and lack of sanitary facilities, with tin baths being commonplace right up until the 1970s. Irene recalls that

> we were very lucky, because my great-grandmother had lived in the house previous to us … and she'd had an extension put on, a big one, where we had a kitchen and a bathroom.

Lynne is still cross with Marian, her old neighbour in Seaton Street, for worsening the poor water pressure in the street:

> The pipes, four houses down, Marian lived there, oh she didn't care about anybody! She just used to put the washing machine on, nobody else could have a washing machine because it wouldn't run, we used to go the wash house off Gilead Street, and we used to go the baths there, but she used to run the water all the time, and she would be out swilling, and we'd be waiting to have a cup of tea!

Peter recalls the close proximity to other residents in Vauxhall Gardens leading to arguments, and family disputes becoming public knowledge; 'The thing to do then when you lived in the squares, there was shouting coming from every house, you didn't feel ashamed because that's how people used to express [themselves]'.

Respondents further recognised that environmental factors also added to the hardships of working-class home making for many. Maureen, Brian, Gary and Teresa all lived in close proximity to factories – bad smells causing Maureen and Brian to remember them negatively – but Teresa has better memories:

> [There were] factories down the back of Great Mersey Street where our living room, we called it the kitchen … and our kitchen looked right out into a factory … it was called, I remember the name because me and my sister used to go and play in the offices there, Curly Springs, and it was just … like oily and dirty, but the fella who ran it, he was a really nice fella, I mean these days you couldn't do it, but we used to go and sit in the office and he'd let us.

Gary, an amputee, feels certain that there is a connection between the amount of industrial and petrochemical activity during his childhood in Speke, and poor health among his peers in the Speke area:

> Sue got her leg amputated, Dave next door got his leg amputated, the woman round the corner got both her legs amputated and then there was me, and you think, just in that small area, that's an awful lot of amputations, there used to be fifteen lads, all knock around and play football and a few of us played rugby, out of that fifteen, nine have died from cancer … . I remember when [daughter] had childhood asthma and I talked to the doctor and he said where do you live? I said Speke, he said [you've] just answered my question…you

> know, you've got Dista, Fords, sending out emissions which some people don't realise, you certainly used to have all the major industries.

Anne and John recall the pests that were ubiquitous to the cellars in the houses of Brasenose Road (Bootle); cockroaches were common, as well as rats:

> Yeah we had a rat man, used to come down our entry, and put the cages down, and every fortnight or week he's come up and pack the cage up, put them all in, empty them in a sack and tie it, and batter it on the wall. (John)

People responded with ingenuity to the problems associated with small, antiquated properties. Teresa remembers

> getting a bath, this sounds really ridiculous, but my mum's brother brought this bath in and they fitted it in the back kitchen and put like a top on, so that was used of a day time, you could just put stuff on it, but of a night time it was a bath.

Similarly, to avoid cooking in the cellar with the cockroaches, as was customary in Brasenose Road, John built a 'kitchenette' in the lean-to at the back of the house, complete with cupboards, a sink unit, and a table and four chairs. The suggestion that 'home' is a verb as much as it is a noun (Mallett, 2004) is applicable here; home is something that we *do* as much as it is something that we *have*, and the 'doing' of home is clearly present in the 'making do', ingenuity and continual managing and negotiating of negative and antiquated aspects of home life.

The postwar urban environment in Liverpool, and the lack of speed in redevelopment, meant that patches of land where properties had been razed to the ground during the war were scattered among the surviving buildings and streets. Respondents old enough to recall the war damage and the debris remembered it fondly rather than as any kind of environmental problem, in keeping with the findings of Balderstone et al. (2014). Several respondents spoke about 'ollers' – hollow spaces left between buildings after wartime bombing – as, rather than an eyesore or a safety hazard, a playground, or a space for a bonfire; 'a great community thing – folk from surrounding streets came to stand around, set off fireworks and roast potatoes and chestnuts in the fire' (Enid). Once 'slum' clearance programmes got under way, the demolition process was also accepted and exploited for amusement; Roy recalls that

> aged five I was with my uncles, going into derelict houses and bombed out buildings looking for World War II artefacts … I quite loved getting out of bed [Seaton Street, Kensington], and looking out the window, and watching the demolition going on, I loved that.

Social divisions also impacted on respondents' early experiences of home. Mary spent the majority of her childhood living in Orrell Park (Bootle/Walton), understood by most as a traditional working-class area; but, because she was born in Portsmouth,[3] and her parents had clear middle-class aspirations, she grew up feeling somewhat conflicted regarding her sense of identity and where she belonged. However, most of the social divisions highlighted by the respondents revolved around sectarianism and the perceived differences and barriers between Roman Catholic and Protestant communities. Jodie, who has a Roman Catholic background, describes the hostility she encountered when she moved into what turned out to be an 'Orange' street (Swanston Avenue, Walton); while Martha, whose father arrived in Liverpool from County Cork, Ireland, aged 14, recalls that 'one of my earliest memories of being a kid was somebody pushing me over, and saying; "I don't want Irish here", it was really odd…and I went home and said to my dad, "I don't want to be Irish"'. Teresa, also a Roman Catholic, recalls that the primary school she attended, Daisy Street School in Kirkdale,

> was like Protestants and Catholics in the same school, but it was split in the middle, it was dead weird because they had their own playground, and we had our own playground, own toilet, didn't mix at all … split the whole school right down the middle, and you didn't even know the people, I didn't know any of the kids who were on that side, and for some reason we didn't mix with the Protestants.

The 'mixing' of Roman Catholics and Protestants was also frowned upon in other aspects of life. Teresa further recalls that her maternal grandfather, who 'was a good Catholic', was forced to marry his 'Orange' wife at the side altar of St Francis de Sales Roman Catholic Church (Walton), as opposed to the central altar normally used for weddings. Gary, of Welsh Methodist stock, was habitually referred to by his Roman Catholic friend's grandmother as 'that Orange bastard'. Christine, another Roman Catholic, worked as a young woman in W Karp & Sons ladieswear manufacturers, and remembers that, while the staff on the factory floor were all 'Orange', and would decorate their sewing machines with orange flowers and crepe paper for July 12, all the women who worked in the offices were, like Christine, Roman Catholic, 'because the Catholic girls were honest and could be trusted'. Christine's testimony, dating back to the late 1960s, is a small challenge to the established knowledge that sectarian divisions in Liverpool were hierarchical, characterised by Protestant dominance and Irish Catholic inferiority (Muchnick, 1970; Belchem and MacRaild, 2006). However, Brasenose Road in Bootle appears to have been an oasis of harmony and mutual respect, as according to Anne and John all the households on the street, including the Glendower Pub around the corner, would celebrate both St Patrick's Day and July 12 regardless of their faith or heritage.

Several of the respondents acknowledged that the childhood or early home that they had loved had many flaws and recognised both that children become accustomed to their surroundings, as they know no different, and that memories often sweeten with distance. This is in keeping with phenomenological conceptions of home which prioritise emotions, experiences and processes over physical space

(Dovey, 1985; Mallett, 2004). Thus, these testimonies indicate that childhood and early homes evoke deeper meanings, beyond the physical state of a dwelling, or the reputation a street or area may have – ranging from feeling 'at home'[4] and a sensation of rootedness (Charleston, 2009) to the psychological notion of home as a transitional object; something to kick against whilst negotiating household rules as one develops into adulthood (Douglas, 1991). As Brink (1995) suggests, childhood homes remain for many respondents their 'real' home where their 'heart' lies, as it is so enmeshed with their relationship with their immediate family and their very identity, to the point where all three become interchangeable. This is evoked, for example, by Jodie as she fondly recounts memories of her father's 'pay day':

> I always remember my dad bringing in my mum, this is no joke right, my dad used to get paid on a Thursday, because my dad used to be an unusual date to everyone else's parents that I knew, so Wednesday was like chips and egg night, and Thursday it was flush night right, and my dad used to bring in a box of Black Magic for my mother and that, coming in from work, I always remember that, and then he'd go Jodie Podie Pudding and Pie, and I used to go, oh here he is and all that, and then he'd give me and our [sister] pocket money.

Childhood and early homes continue to be inscribed on the adult physical body and psyche (Kearon and Leach, 2000). However, loss of family through bereavement can result in the inscription fading and emotional ties to the childhood home breaking, as is the case with Jill, who describes feeling almost as if she no longer *deserved* to stay in the family home once there was no family left:[5]

> obviously with the amount of emotions I felt, and feelings I had for the place, and obviously all the memories, in a way it had changed a bit as well, because everyone had gone I felt it was different in that way, because it was really a family house, and there wasn't a family there anymore … it was just me, and I just felt it changed in that respect, because mum and dad weren't there, and Jack wasn't there, and it just stopped being a family house really … even though I had all those wonderful memories I still felt a little bit, well I don't know if I truly belong here, because I don't have the people around me, I'm not in that family environment anymore.

Thus, the 'authenticity' (Douglas, 1991) of the home was lost for Jill; the house was merely the setting within which family life was conducted (Coolen and Meesters, 2012), and it was the family life that mattered. Family life, for Jill, had turned a space into a place; its loss returned the place into being a space once more.

When asked about their childhood homes and early homes, all the respondents were in concord on a certain point – they all, without any direction or clarification, identified their home at this early stage of their lives as the immediate neighbourhood and community surrounding their dwelling. None spoke solely about their building of residence, and those who spoke of the city centre, or the city on the macro level, contextualised against other cities, did so in relation to recent and contemporary times only. Therefore, the place attachment displayed by the respondents regarding their early homes is very much reflective of *baille* (home town (Mac an Ghaill and Haywood, 2011)) *mahalla* (micro level society, complete with its own distinct set of customs and rules (Porteous and Smith, 2001)) and 'little fatherland' (Ossowski, cited in Lewicka, 2005, p383); concepts which have developed to explain attachment to and prioritising of immediate local district as discussed by Lewicka (2010). The conclusion can be drawn that memories of childhood and early homes are tied here to the local, district level, as opposed to home city level.

People were sorry to lose and tried to keep and protect the home that they had

The lives of several of the participants had been touched, to a greater or lesser degree by the 'slum' clearance programmes of the 1950s, 1960s, and 1970s, and their associated 'modernisation' schedules. Memories of the news that their homes were earmarked for demolition ranged from deep sadness, anger, and a sense of loss, through to relief that the difficulties of day-to-day life in an antiquated living setting would soon be alleviated. The residents of Brasenose Road (Bootle) had lived with the threat of demolition for over ten years by the time it eventually came down around 1970. Anne and John report that they loved living there and 'had it lovely inside' (Anne) although they were precluded from installing an inside bathroom because of the demolition order; "they wouldn't let you put one in, over ten years it was coming down, we went down the Corpy, but they wouldn't let you put one in, because they were wasting their money" (John). The news that the houses were, finally, coming down was upsetting to Anne and John as they had bought their property, had invested in it, and were prepared to invest further had they been given the opportunity. Teresa's family, who also owned their home in Pluto Street (Kirkdale), were equally sad to hear the news of impending demolition:

> I think when they got told they were going they were a bit upset I think because we all lived in the one street, and it was all family all in the one street, they were worried about, my nan and pop, they were in their 60s, probably not much older than I am now, but they seemed really old at the time, and they were worried about us splitting up and being far away from each other.

Irene gave the most clear and coherent account of the loss of her home on Holt Road (Kensington) in the mid-1970s. She had bought the house in the late 1960s with her ex-husband, and to this day it remains her favourite of all the homes she has had. Irene and her neighbours on the same block all fought the compulsory purchase orders on their homes, but, despite an 18-month long campaign, they were ultimately unsuccessful, seemingly due to one somewhat confusing and weak reason:

> We had all sorts of public enquiries, and I've still got the papers actually somewhere, and it said 'no natural light on the landing', that was the deduction of the fellow who come up from London, because we were trying to fight it.

This conclusion is a particular source of upset and frustration for Irene, as it is commonplace for terraced houses to have no source of natural light (a window) on the landing; indeed Irene's current home, rented through Plus Dane Housing Association, has an identical layout to her lost home on Holt Road, including the lack of a window on the landing. Irene has her own take on the demolition:

> What we wanted was to just keep the frontage,[6] and let them build at the back, because they had pulled all the houses down, they were just bulldozer crazy in the 60s, 70s, it was ridiculous, because they pulled all the houses, I mean where I lived in Minto Street they pulled all those down in the 60s, all we were left was a frontage, and then they didn't like that did they, to have this block of terraces which was just a frontage, so they wanted to get rid of us as well.

Irene's continued sense of loss and anger at losing the Holt Road house is consolidated by the fact that it was replaced with what is essentially a grass verge and a wall separating some newer, 1970s housing from the main road which, to Irene, adds insult to injury.

The concept of 'slum' clearance is applied to houses and streets which have been identified as uninhabitable; not fit for people to live in (Muchnick, 1970). However, all of the respondents who were subject to a compulsory purchase order during the 'forward-facing' era of Liverpool regeneration were living happily in their homes, inhabiting them in the usual way. None viewed their home as a 'slum'; rather, all were using the space designated as their dwelling as a true home – eating there, sleeping there, conducting family life there. All cleaned, maintained and looked after their homes; some wanted to do further improvements, but were denied the opportunity. None had *left* the dwelling on the grounds that it was unfit for human habitation. Thus, it seems clear that what has happened in the case of Irene, Anne and John, Lynne and Brian, and Teresa, is domicide – destruction of their homes; not removal of 'slums' that they were no longer able to inhabit. There is a case here then for reframing the demolition of streets like Seaton Street, Pluto Street and Brasenose Road as a deliberate destruction of place, leading to fragmentation of families and communities, rather than a neutral, non-political, necessary programme of clearance of 'slums' which nobody could live in and therefore nobody needed or wanted.

There were however some positives associated with the 'slum' clearances and, other than Irene, memories held by respondents were by no means all bad. Lynne and Brian had hoped to buy their house in Seaton Street (Kensington), and had saved up enough for their deposit, which they ultimately needed to spend to see them through Brian's period of industrial action when he was a bus driver. By the time the bus drivers went back to work, the news had arrived that 'the houses were coming down, so we didn't have a chance to save up the deposit again. But I didn't care, I was getting a garden house!' (Lynne). Thus, the promise of being rehoused in a Council dwelling with a garden softened the blow of loss for Lynne and Brian. Another positive for the respondents was money received from the compulsory purchase order for those who had bought their homes. Irene, although traumatised by losing her home, was glad to receive the £2,500 market value for the property. Teresa's parents

> owned their house in Pluto Street, and they got about £800 which was a lot then, so she [Teresa's mother] was delighted getting £800, I always remember she got like a red carpet and a cream three-piece suite, and like us four we had divan beds, so she had it [their new home] like a little palace.

Similarly, Anne and John, who had been offered a four-bedroom house in Skelmersdale, chose to reject the offer from the Corporation and used their £850 from their Brasenose Road house as a deposit on a three-bedroom semi-detached house in Seaforth. This set in train a process whereby Anne and John bought several homes for their family of six children, each house bigger, newer, and in a more prosperous location than the last.

At the community, neighbourhood level, people were also keen to protect and preserve their homes in the sense of their '*baile*' or '*mahalla*'. None of my respondents had any direct experience of collective resistance to domicide or harm to their community home. However Ronnie, as a former employee of Liverpool Housing Department, was able to elucidate on some of the grass-roots level activism and resistance towards the construction of a second Mersey Tunnel right in the heart of the Scotland Road/Vauxhall area, and opposition to Scotland Road being subsumed into the Inner Ring Road, in which they were successful. The docks being in steady decline by this point, there was less of a pressing need to have a ready supply of manual labourers 'warehoused' close to hand; thus, this dockside community, largely housed in tenement blocks, was increasingly superfluous to the city's changing needs. Significantly, it was the women of the area who resisted destruction of their community landscape, and were the driving force behind the campaigns, much more so than the men:

Clare: My feeling from what I've read is that really it was the women who were instrumental in sort of like campaigning, fighting to protect the area, and so on, and the dynamic ones really were the women. Am I romancing there, or is that fair?

Ronnie: No that's true, and I've found that all over the country, I think it's a bit of a leftover of the men all out at work, but even in recent areas I've worked in, more ex-mining areas, but the men aren't at work anymore, but they're sent to do something; go and do your allotment, go and sit in your shed, so it doesn't upset the equilibrium of the community … the women were the big community leaders.

Clare: That is how it seems.

Ronnie: The unusual thing about Tony McGann and the Eldonians is that it was Tony McGann.[7]

Strong, determined women, and weak, or at least weaker, men, were a continual theme throughout the primary research; a theme which will be picked up again later.

People loved the home that they got

As a result of 'slum' clearance, overcrowding or simply needing a place to live, several of the respondents had experience of either living in, or turning down an option of a dwelling in, many of Liverpool's main 'overspill' or New Town developments; from the earlier ones in Page Moss (Christine), Speke (Gary), Southdene in Kirkby (Pat, Maureen, Colette, Enid) or Fazakerley (Teresa), to the later ones such as Cantril Farm (Lynne and Brian, Roy, Martha, Irene), Netherley (Maureen, Jackie) and Skelmersdale (Anne and John). Some of the offers made to respondents or their families were rejected outright. Martha's parents, who had been living with her great aunt in Whitefield Road (Breckfield), rejected Skelmersdale because it seemed to be a great distance away, and Kirkby, because of her grandmother's bad experience:

> My nan and grandad, my nan only ever went on holiday to Wales in a caravan, never went abroad, never went anywhere but Wales, and they bought a house in Kirkby when it was brand new, and they saved money, and they bought this house, and my nan suffered from depression, because believe it or not it was countryside then, and she couldn't cope, she had been in Breck Road the whole of her life, this to her was like living in the back of beyond, there were no buses, so she was totally isolated when my grandad went to work.

Cantril Farm was their only other option. Irene's parents were only offered Cantril Farm when their Minto Street house was demolished, but they decided instead to look for private rented accommodation in Kensington after paying a visit: 'It was just like a dirt track up to where the maisonette was; my dad said, I'm not moving to the Wild West!'

Teresa recalls that her parents were offered a choice between Speke, Netherley and Kirkby, and, to help with the decision-making process, she remembers taking the bus with her mother and her sister to look around both Netherley and Speke. Her mother was looking forward to getting a newly built house, complete with indoor bathroom – her father, however, insisted on taking a 1930s-built residence in Ladysmith Road, Fazakerley:

> My mum didn't want to go there because it had an outside toilet, she said the bathroom was off the kitchen, it was one of those little ones like that, she said

it's just the same as what we're leaving, but my dad put his foot down, and that's where we ended up … which was a good thing, because it was. It was established, and it was a nice little neighbourhood, it still is actually, I think.

John appears to have similar views about Skelmersdale, where he and Anne had been offered a house, to the views held by Teresa's father about Speke, Netherley and Kirkby:

Anne: Well we didn't want to go to Skelmersdale did we?
John: They offered Skem, and I went up there with Jimmy *****, my mate, and we went there, and got off the bus, I said no way I'm going to live there.
Anne: Because with us having a gang of kids, everyone else would have done, wouldn't they? We were going to a four-bedroom house.
John: Four-bedroom house, and we had to take that one house, and when we were there, there was millions of kids.
Anne: I thought also, when everyone has gangs of kids, that's when they all get in to trouble, don't they, kids.
John: All there was was roundabouts … no chance.
Anne: He come back and he said to me it's terrible, so I don't know whether I got a choice or not really.

When the news came that Seaton Street was to be pulled down, Lynne and Brian initially put their names down for Netherley, as it was slightly more established than Cantril Farm, their other option, and 'Netherley was seen to be the place that people moved [to] from that area [Kensington]' (Brian). However, Brian's job led him to encounter Cantril Farm:

Brian: I was delivering the *Echo*, and I used to come up here, and I said to Lynne; looks quite nice up there.
Lynne: Yeah, I had no idea where it was.
Brian: Anyway, brought Lynne up and we put our name down for, well, it was Cantril Farm then.

Pat, Maureen and Colette's family, however, were so desperate for more space than their tiny home in Wright Street (Scotland Road) could offer all ten[8] of them, that they jumped at the chance of a new home in Kirkby in 1954. In this family's case at least, the allocation of properties appears to have been somewhat haphazard, last minute, and before the area was ready:

Pat: Now the morning we were supposed to, my mum didn't know whether we were moving or not when I was going to work, I worked in Allerton Road then.
Clare: Oh so you had a bit of a jag[9] then?
Pat: Yeah, two buses to Allerton Road, and I said to my mum, 'you'll have to phone me mum, in the office', so she had to go out to the phone to phone

me in the office and tell me, and she rang me, I said 'where have I got to go mum', I said 'have I got to come home', she said 'no girl we're going to Kirkby', came over on a coal wagon up to Kirkby.

Clare: So literally, when you left the house that morning, it wasn't settled?

Pat: No.

Clare: And then your mum rung you and said, 'actually we've got to move'.

Pat: Move to Kirkby.

Colette: On the coal wagon.

Pat: So I got the bus to Lower Lane, and then there was a tram that went along, then I got off at Ribblers Lane, and you had to walk through the farmers' fields; now I went to work, skirt, high heels, black suede.

Clare: I bet you looked lovely Pat.

Pat: Walking home through the farmers' fields, well the mud, and the shoes, and then when I got to our house there was no sidewalk there.

Maureen: It was just wooden planks to get up to the house.

Respondents recalled, in the main, being happy with their new environment. Roy was seven years old when he and his family moved to Cantril Farm, which seemed so very different from what he had known in Kensington:

> We moved up to Cantril Farm, and my bedroom just looked out into woodland which was unbelievable ... catching frogs from a pond in the front ... it was literally a farm, that's why it was called Cantril Farm, Nine Tree Farm that was on the place originally ... and just pure woodlands looking out across farmlands and everything, it was like, wow, never seen anything like it.

Similarly, Martha moved with her parents to Cantril Farm when she was two, and 'according to my mum when they first moved in it was beautiful, it was just like the countryside, and they came from Breck Road, so it was just absolutely gorgeous'. Roy's mother, Lynne, gives an idyllic account of their arrival in Cantril Farm in December 1968:

> The first thing we put up when we got here was the Christmas tree and it was there, that was the Sunday before we moved in wasn't it, and we came up, and put the tree up ... and then the day we moved, I think it was the Thursday wasn't it ... and we come up and it was snowing, and that door was our front door, and our [daughter] was sitting there and I'm saying oh look at the snow, isn't it lovely, we were as warm as toast because we had the gas and the electric put on ... I made the beds, that's all I was interested in, and all the boxes were marked Christmas presents; not to be opened, so they all went in the little room there, and we lived out of boxes more or less for a couple of weeks till I got Christmas over, but it was lovely, absolutely lovely.

The futuristic 'newness' of much of the housing built during the 'forward-facing' period had a varied impact on the respondents. A more 'modern' house was the driving force behind Christine's parents' decision to move the family from their 1930s built semi-detached house in Lordens Road (Page Moss), to a newly built town house in Salerno Drive (Huyton) in 1967; even though the former appears to Christine to be the better quality house. Tower blocks were ultra-modern and, therefore, fashionable – Lynne and Brian both agreed that the three imposing, 22-storey tower blocks at Cantril Farm's shopping centre were beautiful, impressive structures, both inside and out. Ronnie remembers that

> all of the tower blocks sprouted in Everton, I remember going to a wedding in a place I would later work at the base of one of the tower blocks, and just thinking; God I'd love to live here, because it was all new, and everything seemed to be better, so it was the days of the city of change and challenge and progress.

The fashion for building houses on top of houses – two-storey residences on top of others in blocks and towers – was particularly popular in Liverpool, and several of the respondents had experience of living in the 'maisonettes' as they were colloquially known. Irene, on losing her beloved house on Holt Road (Kensington), was rehoused in an early 1960s built maisonette, around the corner in Anglezark Close.[10] Her marriage having broken down by this point, she lived there very happily with her young daughter:

> I liked it. Yeah it was on a landing, you went upstairs, and there were three that side and three that side; I was the very end one that way ... I did like it there, yeah, I did like it there, but you felt safe, it was a safe type of place.

Cantril Farm's maisonettes, known colloquially as the 'Oxo blocks' because of their distinctive cuboid shape, were described by Martha and Lynne as both attractive and desirable; 'they were beautiful inside ... I'll never know from that day to this why they pulled them down, but they pulled them down and they were beautiful, and people didn't want to leave them'. (Lynne)

Maisonettes were not universally popular though. Susie spent much of her childhood in different maisonettes in Beck Grove (Thornton):

> where the new maisonettes were the latest thing ... three houses on top of each other ... imagine living in these concrete, machines for modern living they were called,[11] and that's what they were, they weren't homes, they were just like cold ... if you looked at the estate at that time it was everyone living in these concrete boxes, there wasn't anywhere to play.

Maureen and her daughter Jackie had an even worse experience during the 12 months they spent living in a maisonette in Naylorsfield, Netherley. The strangeness of living in a house on top of other houses, coupled with the final months of Maureen's marriage to Jackie's father, have led to both remembering their time there as very bleak; characterised by poverty due to Jackie's father's cavalier approach to money, paying bills and paying 'housekeeping' to his wife. Both Maureen and Jackie have vivid recollections of hunger, coldness, worry and isolation, all exacerbated by being confined within a drafty maisonette with a shaky, unsafe balcony. Maureen was prescribed anti-depressants to combat the anxiety and panic attacks she had developed and, on the day she left her husband, walked the long journey from Netherley to her parent's home in Fazakerley, with her baby son in his pram and Jackie at her side, because she had no money for bus fare.

Another modernist, futuristic feature of the built environment that proved significant to the respondents was the Radburn[12] system. Applied to the road system in Cantril Farm, Skelmersdale and parts of Kirkby and Netherley, the Radburn system essentially looks to separate pedestrians from motorists and their vehicles (Womersley, 1954). Designed by Clarence Stein and first applied in the 1920s in Radburn, New Jersey, the system was developed to protect pedestrians from road traffic accidents by providing separate walkways, away from the roads, meaning that, in theory at least, pedestrians would be kept out of harm's way (Dumbaugh and Rae, 2010). The Radburn system in Cantril Farm was, at first, a source of bafflement for Lynne and Brian:

> We came up, and we looked, and I always remember a couple of neighbours standing on their steps, and we didn't know how to get into it because this was the back wasn't it? That was the back, and we had to go through the alleyway to get round here, and we were looking, and they showed us how to get here. (Lynne)

All of the respondents who had experience of living in residences built during this 'future-facing' period described an initial sense of surprise and pleasure at the size of their new homes and how much space they had. Enid referred to her new family home in Kirkby as a 'mansion', while Jill recalled running from room to room in her new house in Stamfordham Drive (Garston), delighted with the space, after spending her earlier childhood in a small terrace in Pecksniff Street (Toxteth). Pat and Maureen, who had slept in one small back room with five of their siblings in Wright Street (Scotland Road) were thrilled to only be sharing two to a room in their four-bedroom house in Southdene, Kirkby. Equally as valuable was the hot and cold running water and, most importantly, the indoor toilet:

> which absolutely amazed me when we first went to see the house, I just remember sitting on the toilet, just sitting on it and actually amazed that it had an indoor, because I'd never really seen an indoor toilet up to that point ... I was just absolutely wow in amazement at this indoor loo, and an indoor bath, an indoor bathroom. (Jill)

After having difficulties with the low water pressure in Seaton Street, worsened by Marian's washing machine, Lynne and Brian took a while to get used to the strength of the water pressure in their new home; 'we kept breaking cups, the pressure of the cup under the water, bang, honestly the cups would just go' (Lynne):

> my mother, when she moved up to [Cantril Farm] used to turn the tap on and smash cups because the force was that bad, literally used to drag it out [of her hand], and we were getting through a cup every couple of days, shrieks of laughter everywhere, holding the handle. (Roy)

Other utilities such as gas and electric were also a boon. Lynne and Brian's daughter was aged two when they arrived in Cantril Farm, and the piped warm air to heat the house was very welcome in December. Thus, the modern conveniences that houses built during this period offered their inhabitants were life changing and life enhancing. Yet the facilities were not always enough to ensure happiness within the home, as Susie's testimony indicates:

> At first everyone was so happy because most people who come there come from the slum clearances, so they, as I say, running water, bathroom, three bedrooms, all this modern stuff, they were chuffed, but Mum got disillusioned really quickly and was like, I don't like it, I want to get away, and she wanted us to get away because she said it's not right, there's nowhere to play, and all that kind of thing.

In spite of Susie's testimony, which indicates that her mum detected a lack of authenticity in her new home, much of the testimony of the respondents runs contra to the established knowledge that 'slum' clearance was inherently bad for communities, and that, for example, peripheral housing estates were worse places to live than the places that people had lost (Muchnick, 1970; Couch, 2003; Murden, 2006; Rogers, 2012). Newly built residences, whether they be in established communities, in overspill developments, or in new towns, offered much to their new residents – from better and more reliable utility services to increased space and access to green areas – than antiquated inner area housing could. The 'fresh start' that a brand-new house, especially in a newly established setting, gave people and families acted as a springboard for many to greater ambition, prosperity and personal fulfilment, jump starting an increased sense of optimism for the future. Thus, while developments like Cantril Farm and Netherley were by no means perfect, they were by no means fatally flawed either; arguably, increased and continual maintenance and monitoring of such areas in their first decade of existence may have negated the physical, social, and reputational problems that they began to develop from the late 1970s onwards. It is important to acknowledge that families subjected to 'slum' clearance are not passive victims of a social process over which they had no control; rather, they are better viewed as strong and resourceful people who, while perhaps experiencing sensations of loss regarding their old home, simultaneously embraced the future offered by their new home and negotiated difficulties with resilience, confidence, and positivity.

Difficulties and setbacks to overcome

Whilst it feels vital to stress and emphasise the positive aspects of life in newer versions of the 'Liverpool home', as uncovered in the testimonies of respondents, it is equally crucial to acknowledge the threats posed to home and family life by such developments, and the resultant harms to the lives, relationships and general well-being of the respondents. Muchnick, writing in 1970, feared that the inner area and peripheral overspill developments would, through poor or non-existent planning, simply result in the recreation of the 'slums' the housing department sought to eradicate; from testimonies, and taking into account the various structural problems that arose at the same time, it would appear that he was, at least in some cases, right. Many of the people who shared their stories with me did encounter a range of difficult life events, frequently related to processes impacting on their immediate home (dwelling) life, the prosperity of their neighbourhood and the economic, political, social, and reputational processes occurring at the city level.

Promises made by the housing department to assign residents of cleared 'slums' to the same new housing development, thereby preserving existing community links, were often not kept (Broomfield, 1971). Many members of Teresa's family had lived close to each other in Pluto Street (Kirkdale); once the street was demolished, the extended family were scattered throughout the north end of the city. Teresa's grandparents, together with her youngest aunt who still lived with them, were particularly keen to move close to Teresa's parents in Fazakerley; the closest they could get was a 9th-floor flat in a newly built tower – Storrington Heys (Croxteth). For Teresa, the move at the very least contributed to her grandfather's death soon after:

> I think he was only in there about 2 years and he died, he was only 65, and I think, because he used to go to St Alphonsus church [Kirkdale], and he was sort of running it, and he'd go to the club afterwards and things like that, and he used to have a bad chest, so walking up Storrington Heys, because my nan said she used to see him walking up and he'd have to stop, and he'd be wheezing to get the bus, but he still couldn't go to the church, Queen of Martyrs, round the corner, he had to go down there, and yeah, it was really sad.

Lack of appropriate town planning, as opposed to housing provision, meant that newly built residential areas were often under-developed (Muchnick, 1970; Couch, 2003; Murden, 2006). The initial lack of local amenities, and good public transport links, added to the difficulties of settling into a newly established area. Anne and John had no real shopping facilities near their newly built private property in a development in Aintree, a semi-rural location where 'they were growing wheat right up to our back-yard door' (Anne). Similarly, Lynne and Brian had to wait for shops to open in Cantril Farm and when they did, they were fairly limited in nature. Public transport was also a problem – when Lynne and Brian first moved to Cantril Farm, 'the buses only ran as far as the Princess pub [West Derby]'. By the time Martha and her parents arrived in 1970, the bus links had improved; however, there was no nursery for two-year-old Martha to go to while her mother went to work as a wirer in Plessey's.[13] Thus, Martha's mother had the daily task of taking her daughter by bus to a nursery in Everton, then walking two miles to the Plessey's site, and then repeating the journey in reverse at the end of the working day. This resulted in her mother ultimately giving up her job, in spite of the fact that she was the chief earner in the family.[14]

Many of the respondents had experience of poverty, in some cases abject poverty. Jodie recalled stories she had been told by her parents about their early married life before she was born, when they only had her older sister:

> They moved [to a flat] on top of a launderette, and my dad said, like everything used to be money then, coins, so my dad said when they were skint all that they used to do was wait for someone to come and rob the machines, because they were always leaving loads of money all over the floor … . I think financially, because they were only kids, and when you start you've got nothing, have you, and my dad says; we're sitting in the flat and we had one cup, *one cup*, and my dad said to us it was the happiest time of our life, one cup, that was it.

The furnishings that Pat's family had in their small house in Wright Street looked sparse and scant in their new house in Kirkby, and it took some years before the house was fully furnished:

> Me and our Margaret said, because we were courting then, she was courting, I was courting, and we wanted the parlour looking nice, so my mum got a rug, I think it was a two by two rug, and me and our Margaret used to give her half a crown a week to pay it off, and she'd pay it off because she got it on hire purchase … there was one wardrobe in the house, which was built in, wasn't it, in our Paul's room, just a little bit with a rail, that was what we had, and at the side of our bed our Margaret and I had like an orange box, you know, with like a shelf in between, and that's where we put our clothes, nowhere else to.

Cutlery and crockery were also scarce in the family of nine children, so much so that, when the children wanted to swap an item with the local rag and bone man, their mother made them choose a plate to augment the family's crockery supply:

> Colette: We used to run out and we used to take our clothes out to the rag man, and I said, I ran out this day and took a coat, I said I'm going to get a balloon, that's all I wanted, a balloon, and my mum used to say you're not getting a balloon, you're getting a plate.[15]

Although most of the respondents had very working-class backgrounds, and many had endured some level of hardship, Jackie's childhood with her brother and her mother Maureen was characterised by extreme poverty, including deprivation, fuel poverty, and hunger. Jackie's late father, Jimmy, was a drinker and a gambler, and it was a regular occurrence that Maureen would be left to manage on the Family Allowance she was entitled to for her two children. Weeks would pass without Jimmy handing over any of his wages from his job working at the dock gate, and Maureen would struggle to feed herself and the children, relying on Oxo beef stock cubes as they were both nutritious and cheap; 'I used to have nothing in my cupboards, Oxos, that's all, that's what I used to call my kids ... the Oxo kids, used to go to bed on Oxos and butties'. Maureen described how she would wait anxiously for Jimmy to come home:

> I used to stand by the window upstairs, you [Jackie] didn't know that when we lived in Naylorsfield, I would be at the window like that at 1 o'clock in the morning, saying, oh God where is he, in case he comes in shouting, I would go and stand by the door, frightened again, I don't know how we lived there Jackie, stood by the door waiting for him to come in, and I used to say Jimmy are you going to bed, I was shaking like that, worrying, he'd been out drinking and my kids were in bed with Oxos [in their stomachs], nothing to eat.

Jimmy's failure to pay utility bills resulted in the gas and electricity supplies to the residence being cut off, and Maureen struck a deal with her next door neighbour, an elderly man – she would clean his house for him, and in return he allowed her to cook hot food for her children in his kitchen.

Jackie was a young child during this time and, as such, was sheltered from many of the effects of poverty. However, she does recall developing strategies for procuring treats that her mother was unable to afford; Jackie noted that one of the neighbouring households would have regular visits from family members, including children and, on such occasions, a tin of biscuits would be brought out – Jackie learned that if she sat in the stairwell during such visits, she would be invited to play with the children and offered a biscuit. Similarly, she worked out that her father would only say yes to her requests for money for sweets when members of his family were around; on all other occasions he would say no. Perhaps her bleakest recollection is waiting outside the local pub, The Bridge, with her brother in his pram, whilst her mother went inside to plead with her father to come home or, at least, give her some money for food shopping.

Several respondents, including Maureen, spoke of asking for 'tick' or 'trust' in shops in the hope of taking what they needed and paying later. Similarly, other methods were used for spreading the cost of essential items such as clothes and household appliances – including 'catalogues', money lenders and the so-called 'provy' or 'provo' cheque – a personal loan issued by Friends Provident which could only be spent at certain shops. Pat and her sisters Maureen and Colette, Enid, Maureen and Susie all remember reliance on the 'provy' cheque for buying household essentials such as clothes and small items of furniture. Maureen, for example, would use the cheque to ensure that Jackie and her brother had new clothes, including underwear and shoes, every Christmas. However, there was a certain level of stigma attached to the 'provy' cheque, and making purchases from the 'catalogue' – essentially a brochure for a mail order company whereby payments could be spread over a period of time – was seen as both more manageable and more respectable:

Martha: We got all our clothes from the catalogue, my poor mum used to have bills and we always paid it, everything was from a catalogue.
Clare: Did she used to have provy cheques, or anything like that?
Martha: No! No, she wouldn't have a provy cheque; she used to think that provy checks were like, once you started on the Provident, that was like the end of it, but she didn't mind Kays Catalogue, and she'd pay off, it never ended, once one thing is paid off, get all my bills down, but if you spoke to my mum she would never admit that she ever cried over a loaf of bread.

Susie's mother would buy clothes from a certain shop 'on the drip' i.e., by paying in weekly instalments:

> Christmas time we'd go to Mrs Acky's shop, and you could get an outfit, and me and my sister would get a new outfit, new shoes and a new coat, and then my mum would be paying it off then, and then the next time would be June because we both had our birthdays in June, and we'd get another one.

On occasion, Susie's mother would get into difficulty with her repayments and items would be repossessed:

> Oh God yeah, we were in hock to everyone, we come home [one day] and the washing machine was gone; I went, 'where's the washing machine?' It was one of them twin tub ones, you know, she said; 'oh I can't be bothered with it, it gets in the bloody way, so I told them to take it away', well she didn't, she just couldn't afford to pay £2 a week, or whatever it was to the man who would come.

What is important to note here is that these individualised experiences of poverty did not occur in a vacuum – they are reflective of both the determining contexts and social mores of the time, and the structural problems impacting on the city as a whole during the 1970s. Carl explained the reasons behind his decision to go away to work in Jersey as a young man in the early 1970s:

> There was no money, there was absolutely nothing, like I said there was no fuel because we had the petrol strike, you had the miners' strike and there was just nothing, and like you say you couldn't go home and watch the telly at night because of the blackouts … it was bleak, it just went, no heating, seven o'clock people would be going to bed.

Teresa has similar memories from later in the decade:

> All the big factories were closing down like the one where I worked, Tate & Lyle, all down there, they were all families who worked there, and it was all families that were out of work altogether, my dad got finished in Tate's, he was a lorry driver, and he'd worked all his life; we weren't kids when my dad got finished, but I can remember my dad sort of just lying on the couch, and he wasn't that type of man, he must have been so depressed, he wasn't getting shaved and my mum went he'll be alright, that generation, they never really talked about it, but I think it was a sad time because it was just awful.

Thus, while the poverty experienced on an individual level by Maureen and Jackie was caused by Jimmy's actions, it must also be noted that it coincided with widespread poverty throughout the city and was exacerbated by the gender politics of the day.[16] As a married couple with children, Maureen's role was caregiver and homemaker whereas Jimmy was expected to provide for his family; Jimmy's failure to fulfil these expectations left Maureen trapped, with limited opportunity to improve the family finances and a limited voice in trying to get support from Jimmy's family to encourage him to do the right thing.

The same gendered poverty affected other respondents too. Susie's father died when she was nine, leaving her mother widowed with two children; as a nurse she struggled to make ends meet on her single salary, and was 'too proud' to claim benefits. Martha's mother, who had given up work when she was expecting her second child due to lack of support with child care from her husband, was plunged into poverty when her husband's sudden health problems coincided with structural factors outside of their control:

Martha: My dad was involved in quite a nasty accident, he was always fit my dad, and he was on his bike, going down Rocky Lane, and someone opened a [car] door and literally he went right over the door on his bike, so then he became unemployed, my dad.

Clare: Because he was disabled?

Martha: Yeah, so he had an operation on his back, so we went from having lots of money to basic … he got better, then he got a job working for Barratt's [Homes], my dad would always, even when he had the job in Plessey's, he was actively involved in the unions, so he was a union member, and then he became a shop steward, and Barratt's, they were on strike, and my dad was a shop steward then, and they were out, donkey jackets on, sitting round a fire, and he was out for I'd say about 4 months, but he got blacklisted then.

Clare: Because he was with the union?

Martha: Because he was with the union … but he never worked then, so the rest of my childhood with my dad on benefits, and then, because he was handy, getting scraps of work, he used to, like rich people in Childwall would get him to do houses, and he'd get like £20 a day, but he used to come home, my dad

> was lovely, he used to say; 'I've put up paper that was £20 a roll', because he was amazed at the riches they had, and he was getting like £20 in his hand for doing it. So we were very poor, yeah.

Thus, the gender dynamics that kept Martha's mother out of work, and had similar effects on Maureen and Susie's mother, colluded with class dynamics with the result that Martha, her parents and her two sisters existed in poverty for several years.

Arguably these women have been subjected to a gendered construction of place – the notion that a woman's place is in the home has rendered the experience of home as both different from, and worse than, that of men; from Martha's grandmother developing clinical depression due to her 'exile' in Kirkby while her husband was out at work, to her mother's enforced poverty through her father's refusal to share the burden of child care, and Maureen's quasi incarceration in a cold, drafty, almost empty maisonette, with no money and very little food, resulting from societal expectations that her husband will 'provide' for her and the children. In this sense, these women were *placed*, and *held in place*, often against their will, as a result of their status as women, and working-class women at that. Thus, the rest and relaxation that 'home' is supposed to provide is not always available to women, and the privacy of the home can also mask its harms.

Whilst gendered poverty has clearly had an impact on women respondents, reflecting the social dynamics of the day, male respondents were also subjected to negative processes arising out of a coalition of class and gender oppression. Peter notes that his coming of age coincided with both the continued decline of Liverpool's economy and Thatcher's ascent to power; 'when I left school it was, actually unfortunately when I was 18 was when Margaret Thatcher came to power, and jobs were very hard to come by at the time'. In noting this, Peter displays his knowledge and understanding that processes occurring at the structural level were limiting his life chances and his opportunities to embody and perform working class masculinity:

> I was unemployed … well I say unemployed, but I'd done various training, well it used to be called the ET, used to call it the 'extra tenner',[17] but with me I tried to do things for the sake of it rather than getting harassed off them, because the way it is now, the modern film, I Daniel Blake, it was creeping in then, my stomach used to turn, and I'd get made to feel like horrible, and it was no good for me, and I think that was the early start of me getting into depression and thoughts like that into me because the way I was made to feel … it was a horrible life, there would be nothing I would have loved more than a proper job, but they didn't give us a proper job, they were trying to give us what I would [call] slave labour.

Significantly, Peter has an awareness of his location in the class structure, and considers himself to be knowledgeable and self-aware, in the Freirean sense;

using his awareness of his structural positioning as a mechanism of both resistance and re-affirming his working-class masculine identity:

Peter: My education started when I left school, but on my mam's side, my mam was very political, and my uncles, they were very political, from a very political family, I'd say like as much as my mam was into politics ... but I learned more off [uncle] Brian...I'd see Brian on a Wednesday for about 5 years, we'd go out on a Wednesday night just me and Brian, and then sometimes on a Friday it would be me, Brian and my Uncle Tony, so he was like a big influence on me politically, and then obviously with Tony as well, the two of them together, and we'd always be talking about *Newsnight* and that, columns in the [Daily] *Mirror* and things like that, and I was learning all this, taking it all in.

Clare: So *you know* that a lot of what's gone on, like the things that have happened in your life.

Peter: Yeah.

Clare: They're nothing to do with Peter ****** really, because you're just another fella.

Peter: Under the circumstances, yeah, I know what you're saying, and I agree totally with what you're saying, not just me, thousands of other people.

Thus, Peter recognises that his circumstances mirror those of a generation of men whose life chances were seriously curtailed by a convergence of processes under Margaret Thatcher's regime (Dorling, 2006).

Poor education was one of the causal factors in Peter's poverty. Peter went to a secondary school in the Scotland Road area called

> Archbishop Whiteside, it was the worst achieving school in the whole of Britain, and the reason I know that is because I'd seen a documentary on it, the worst achieving school in the whole of Britain, which it was, it was a school where anything was more or less allowed, and no one got an education unless they were already very clever ... the rest were left just throwing things at each other, and it was just a bad school.

Reflecting the 'stupid and thick' discourse that had developed about the inhabitants of Scotland Road (Courtman, 2007, p205), Peter identifies structural factors pertaining to the status and reputation of the area as the cause of the school being so weak:

> There were good teachers, but they were sort of fighting a losing battle because I think the whole infrastructure and the whole thing was like; oh all them go there, you'll get them all there, basically everyone who went to Archbishop Whiteside was everyone from all the junior schools all dotted around that area, the Liverpool 3, Liverpool 5 area, and it was just oh get them all in there, and it really was poor standards, and looking back yeah I

> think the system and the powers that be, that's the way it was to me on reflection, most definitely.

Cantril High School in Cantril Farm, the only secondary school on the estate, was attended by both Roy (1970s) and Martha (1970s/1980s) and appears to have suffered similar problems. Roy first attended a school in Cantril Farm aged seven – Nine Trees Primary – and found it very different from his previous school, Clint Road School in Kensington:

> The biggest thing with Clint Road was the discipline, the discipline was absolutely regimented, I remember, the nearest I can say to you is Hogwarts, if you can just imagine Hogwarts, and it's all spiral staircases here there and everywhere, and it was just amazing … . Nine Trees itself wasn't bad, it was a new way of thinking, it was nice new young teachers, all very 60s, all huggy, yeah, let's get the best out of [the children] … even then I didn't think the education was as good as Clint Road … . I remember getting hit on the back of the knuckles by teachers in the junior school at Cantril Farm for using my left hand, but I'd been doing double writing, italic writing and copper script up at Clint Road with fountain pens and everything age 6, 7, and [at Nine Trees] I was told not to do joined up writing, told not to use a pen, you're only allowed to use pencils, and don't dare write with your left hand.

But for Roy a bad situation was made worse when he went on to Cantril High:

> The real down turn came at secondary school … absolute worst place on the godforsaken planet … should have been bombed off the face of the earth the year that it opened … it seemed like an experiment, a sort of social exercise, it was bizarre, it was like you had teachers, most of whom were NQTs [newly qualified teachers], anyone who was anything more than an NQT usually ended up in a management position, so you had all these people fresh out of teacher training college, and they didn't have a clue, and you could run rings around them, it was like a zoo. The people in the upper sets tended to escape the people in the bottom sets, but you still had to mingle with them at break times when the bullying went on.

Martha's experience of Cantril High a few years later was similar:

> I started senior school, that's when it became tough to live on Cantril Farm, I didn't fit in, my dad was very strict, he was a strict Irishman, you went to school, you worked hard, there were rules in the house, you sat at the table, you'd say 'please may I leave the table', and that didn't fit in with the way of life in Cantril High … going to school in Cantril Farm was not about going to learn an education, it was about messing around, and you just couldn't get an education, I used to have teachers, well, teachers didn't used to care, they used to smoke in the classroom, they used to go into the staffroom and have

> bottles of whisky … when I think back now, it was a bit like hell, because you were living in an estate which basically was run down, there was a lot of poverty, and then you were going to school, and you had teachers who were mad, who basically were giving you a good hiding.[18]

Unlike Peter, both Roy and Martha managed to do well at school, and both left with qualifications. Roy achieved the required five O levels he needed to join the Royal Air Force as a radar technician, while Martha left with O levels in English, English Literature, and History, plus several Certificates of Secondary Education (CSEs). However, Martha feels she was badly let down by the school:

> This is what's quite sad, I was in the top class for everything, but the teaching was just awful … when I think about it now I was a hard worker, I was in every lesson, my dad would never let me stay off, but the teaching just wasn't there, I don't think the teachers were interested in us, it was hard with large classes, and the majority of them were basically just messing about.[19]

In spite of this, Martha was accepted into tertiary college; 'I was on a lovely path, I was doing A-level English Literature, A-level Law, and A-level American History'. However, a Thatcherite policy resulted in Martha voluntarily giving up two of her A-level programmes,

> Margaret Thatcher brought in a policy whereby if anyone was in the house, and they were working, they had to contribute, well I was at college, and I was getting a grant, and they literally stopped all my dad's money because they said that I was an adult now, and I remember having to go, and I thought it was demoralising, I had to go with my dad … and beg with him for money which was awful really. It really was, we had no money, it was Christmas time, and we just had nothing because they had stopped all benefits.

Thus, Martha left full-time education;

> We couldn't afford to do it, my mum and dad would never admit that, but we just never had any money, so I dropped 2 A levels and kept the Law, because I really wanted to do the Law, and I [studied] part time over 2 years … I just said to them 'I can't cope with the A levels' … so it was like it was me that wanted to do that … it's just the way it is when you're working class, you've just got to do it, you don't really get a choice.

Taking this decision meant that Martha was able to take a job as a legal secretary with a wage of £50 a week; most of which she gave to her mother after paying for a weekly bus pass to get her to and from work.[20]

Roy differs from Martha and Peter in that his family had never really known poverty as such – his father Brian always worked, and his mother Lynne supplemented the family income with part-time work where possible. This might explain why he has a different take from Martha and Peter on the problems that beset Cantril Farm, and areas like it throughout Liverpool, in the late 1970s, as indicated in this excerpt:

Clare: Do you think that there were any other factors influencing a high crime rate [in the area]?

Roy: Lack of discipline and deterrence.

Clare: Really?

Roy: Simple as that … it was one of them, oh well, if you get caught nothing happens to you, you get a slap on the wrist, that's it, and people were able to get things they weren't able to afford otherwise so why not, why flog yourself to death, when you can go and pinch something for nothing?

Clare: Do you think, because obviously the thing that strikes me, the change … round about the late 1970s, now that screams at me Thatcher, and Thatcher policy, and sort of an economic downturn in the area and less jobs for young lads. Do you think, is that something you can relate to?

Roy: No, I don't think that at all, I'm not a Thatcherite … I think sometimes she can be a bit of a scapegoat, and she can be used to justify a lot of things that weren't actually her fault, I think a combined thing of lack of discipline, yeah ok there was not as many jobs, but what stopped people going out and starting businesses themselves … yeah, people were very downtrodden [but] I'd nail nearly every problem around that time down to education, and lack of foresight basically as well as it becoming increasingly easy to earn a living via underground means, you know, why go and do something legitimately when 30% of my earnings went in tax?

Thus, while both Martha and Peter are able to contextualise their experiences and their reduced opportunities against a backdrop of both local and global processes impacting on their structural location, Roy takes a more authoritarian populist stance by conflating entrenched social and economic deprivation with individual morality, discipline, and work ethic (or lack thereof). Martha and Peter both recognise, in the Freirean sense, that their stories are also the stories of many other oppressed working-class people across the city, the country, and the world; Roy, however, focuses on individual decisions and life choices rather than structural constraints.

The contrasting viewpoints of Martha and Peter on the one hand, and Roy on the other, can further be examined in relation to notions of space and place. As noted in Chapter 1, the term 'house' acts as both a noun and a verb; many of the respondents have been passively 'housed', often with very little input, by the various housing bodies that have operated in the Liverpool area throughout the years. Where people have been housed, and the manner in which they have been housed – from tenement blocks to parlour houses – both reflects and influences their structural location; their *status*, whether in terms of their perceived social class, their ethnicity, or their religious identity. The testimony of Martha and

Peter clearly demonstrates knowledge and awareness of their structural location; a position which was imposed on them and against which they had limited ability to resist. In this sense, both Martha and Peter demonstrate a sensation of being *placed*; of being put in their place, kept in their place, forced to know their place, with very little opportunity to escape their place. Their 'place' is both metaphorical, in the sense of their location in a hierarchical class structure, and physical, in terms of the place where they live, their limited opportunities in the place that they live and, significantly, the reputations attached to the place that they live (Tuan, 1977; Waghorn, 2009; Withers, 2009).

Conversely, Roy's testimony on his life in Cantril Farm appears to indicate a sensation of *space*; having the freedom to take up opportunities and the ability to resist structural forces conspiring to keep young men in the 1970s and 1980s in Cantril Farm, Scotland Road, and other areas like them, in their *place*. This may simply be a result of the fact that Roy has no direct experience of the hardships of poverty in his own family life. It could however also be the result of him joining the Royal Air Force at a young age, providing him with literal freedom to travel and acting as a force for equalisation:

> You get people from Newcastle, Glasgow, all of us, you're just a sort of melting pot in the forces, you'll have a universal forces accent, but everyone got on with everybody, there was no, on the squadron, hey Scouse, hey Geordie, whatever ... in a good way, very much so.

A final difficulty that respondents raised in connection with this time period was crime, perceptions of increased criminal victimisation, and the emergence of a criminal reputation. For Lynne, Brian and Roy, the worsening crime picture in Cantril Farm in the late 1970s was exacerbated by aspects of the built environment in the area, specifically the Radburn system, which facilitated easy escape from and evasion of the police. Narrow pedestrian walkways, often covered, were known as "rat runs" (Lynne), woods surrounding the estate offered seclusion, and underpasses and subways promoted antisocial behaviour in the form of graffiti, substance misuse, urination, and defecation. Drug use and its effects were a particular problem impacting on Jodie's home life with her family in Walton:

> That side [of the estate] just went crazy ... they were more into the drugs, there was all different things going on, and everything like that, at the time you could rob cars, rob radios out the cars, that was the big main thing then, it just went rough, and then people started robbing off each other, because other people were coming into the street, because they were wanting drugs and everything, next door to my mum's there was always addicts there, and obviously people were looking at my mum and dad as though they had money, because we were going on holiday, and my mum had the best gear, she was spotless and all that, and it wasn't that they had money, it's just they went out and worked, and spent the money on the likes of clothes and holidays.

Teresa describes the impact that drug use had within her own extended family:

> My cousin, my first cousin, we're just from a normal family who none of us have ever been in trouble or anything like that, and when he was 16, think he's 48 now, it was a typical like one in every family wasn't it, like with the heroin, and he just ruined his own life, his mum and dad's life, he's still ruining their life, they're bringing up his child.

Respondents, therefore, were able to identify a range of criminal and antisocial behaviours as reaching a climax in the late 1970s and early 1980s, corresponding with the 'perfect storm' of industrial and economic decline, increased heroin addiction and football-related violence established in the literature (Topping and Smith, 1977; Fazey, 1988; Pearson, 1991; Thornton, 2003). The cumulative effect of these processes on Liverpool and its inhabitants is encapsulated for Teresa in *Boys from the Blackstuff*:

> It seemed grim didn't it, it was all like the working class seemed to just be poorer, and I would never count myself as being poor, but we were always working class, we didn't have big fabulous holidays, we went to Butlins every year and all that malarkey, but everyone seemed to be dragged down, and I think that's what I was saying about Boys from the Blackstuff, they got it so well didn't they.

Significantly, many of the respondents identified this period as the birth of the negative, criminal reputation that continues to blight Liverpool and Liverpudlians. Almost all of those interviewed had direct personal experience of derogatory stereotypes, particularly pertaining to crime:

Like the stereotypical thing, you're a robber, you're a thief, you're dishonest and all that carry on. (Tony)
They think we're all scallywags and rogues don't they, loud mouths. (Rita)
It is a city full of hard knocks and scallywags, a city full of thieves. (Joey)

This well-recognised and well-documented criminal and antisocial reputation is resented by respondents, not least Maureen, whose son Gary was unlawfully killed at the Hillsborough disaster in 1989. Maureen and her sisters Pat and Colette feel certain that reputations attributed to the city of Liverpool generally, and Liverpool Football Club supporters specifically, were clear causal factors in both the disaster itself, the result of poor decisions taken by senior police officers, and the way that the disaster was subsequently responded to by the emergency services, the Tory government of the time, the media, and, ultimately, the wider world:

Pat: Our two lads were in the police and they felt absolutely hopeless, ashamed really of what was going on.

Clare: Yeah, it's like, I mean obviously we all know now, but it's like the rest of the country were looking at us and trying to make us feel ashamed.
Pat: Like hooligans, Margaret Thatcher.
Colette: You all caused the trouble yourselves.
Maureen: For her it was the icing on the cake, oh Liverpool again.
Pat: Threw everything at them.
Maureen: She'd been involved in putting the fences up anyway, and those fences, they still came in like that.
Pat: They were like barbed wire weren't they.
Clare: Yeah, deliberate, spikes to stop people climbing and all of that.
Maureen: As Phil[21] said last night, he showed us like a few reports and one of the words that they used was 'corralled', and he said, 'what do you corral?' Animals, cattle, and they said they were in pens.[22]
Clare: You were talking about pens weren't you?
Pat: Yeah, cattle.

Further, as well as the stereotypes of Liverpudlians as a whole, there is also an 'internal' system of reputations whereby some districts are viewed with disdain by the residents of others; Cantril Farm – colloquially still referred to by some as 'Cannibal' Farm[23] – being an area subject to particular contempt, as illustrated by both Martha and Roy:

> My first boyfriend, who I'm still in touch with, actually [the] first time I met his mum and dad, and he said 'this is Martha', and they said 'oh where are you from', and I said 'I'm from Cantril Farm', and you should have seen their faces…in fact I remember his mum who was lovely saying to a friend; 'we're worried you know, he's started going out with this girl from Cantril Farm'. (Martha)
>
> Ex-girlfriend of mine, when her parents knew I was from Cantril Farm, oh my God, literally, keep the car keys hidden, it's like excuse me, and to a point where it did affect our relationship, and we did end up splitting up. (Roy)

These stereotypes and reputations cast a long shadow over the respondents, who universally see them as both largely unjustified and, in cases where there *is* potential justification, outdated. With the possible exception of Roy, the respondents recognise that such reputations emerged during a period when Liverpool and its environs were suffering from a variety of negative structural processes and, crucially, it is these structural factors that have caused individual responses, sometimes criminal, which have given rise to the negative stereotype of the workshy, dishonest, criminal Liverpudlian. Thus, the unique confluence of various punishing negative factors that have blighted the city has created the unique regional stereotypes that remain detrimental to Liverpool's reputation, and hurtful and unfair to Liverpool people.

The next chapter, Chapter Seven, presents the remainder of the data gathered and analysed, following the same thematic framework and starting with the demise of the forward-facing era and the simultaneous emergence of the backward-facing era.

Notes

1 Peter, for example, spoke fondly of his early life in Vauxhall Gardens, but acknowledged there were unhappy times, particularly in the lead up to his parents' marital breakdown.
2 Jill's father, refusing to accept the medical opinion that his son was 'a cabbage', spotted an article in *The Observer* about an institute in America which helped children with brain injuries and developmental needs. Building on the groundwork done in Philadelphia, paid for in part with the 'whip round', Jill's father devoted his life to helping Jack develop his intellect, teaching him to read and write, and encouraging him to enjoy the written word.
3 Mary's father was in the Royal Navy and, as a result, the family lived in a variety of naval towns before returning to their native Liverpool. Thus, Mary was born in Portsmouth and her older brother was born in Haverfordwest, South Wales. This, coupled with her mother's world view, meant that Mary was encouraged to see herself as 'not Scouse': 'I think with my mum it was a certain amount of snobbery…my mum saw herself as a little bit above, and she saw Scousers as, you know, people who lived on Scotty Road and all the jewellery, wore all their wealth on their hands…and she *wasn't* that, so she wouldn't align herself to that'.
4 Jodie, one of the youngest respondents at age 39, describes a particular type of relaxation that she only feels when she visits her parents' home, where she grew up; 'my mum's [house] I associate with relaxing, I love going over to my mum's, and I could just get on the couch, even sometimes I just love going into their bed, and I could feel I could just go into a nice sleep when I'm in my mum's'.
5 Jill spent the majority of her childhood and adult life in a large 1920s three-bedroom Council property on Stamfordham Drive, Garston; the death of her parents resulted in her brother Jack going into supported accommodation, meaning that Jill was left alone in the house.
6 Holt Road is a relatively busy main road which has several terraced streets running off it; Irene's house was in front of a series of terraced streets, including Seaton Street, where Rita, Lynne, Brian and Roy had all lived, which were demolished in the 1960s. By 'frontage', Irene means the houses that faced onto Holt Road.
7 The Eldonian Village in Vauxhall arose from a housing cooperative, effectively 'led' by a man, local resident Tony McGann, which formed in response to the demolition of tenement blocks in the area. Keen to remain in the area and close to existing neighbours, the cooperative procured the derelict former Tate & Lyle site for housing for displaced local families (McBane, 2008).
8 Pat, Maureen and Colette and five of their siblings were born in Wright Street; their youngest sister was the only one to be born in Kirkby.
9 'Jag' being a colloquial term for a long, tedious journey.
10 Irene stayed at this place of residence until 1996, and only left because the building had become structurally unsound; subsidence, caused by water coming out of the nearby reservoir, had resulted in gaps appearing between the ceilings and the walls. Thus, the maisonettes were demolished, causing Irene to lament that 'every time they moved me, they pulled my house down'.
11 Susie appears to be making reference to the notion of houses as 'machines for living in' associated with the work of architect Le Corbusier (cited in Lund, 2011, p59).
12 The Radburn system, or the Radburn-type system as it is better called in relation to the UK examples, was used widely throughout the 1960s – however, it was often modified to cut costs, particularly in the case of Council estates (Ravetz, 2001). This sort

of corner cutting, together with the problems that often occur when families living in poverty are ghettoised (see e.g., Hanley, 2007), have given the system something of an unwarranted bad press (Lee and Stabin-Naismith, 2001).

13 The Plessey Company was part of a 'hub' of telecommunications industry in Liverpool for over 100 years, employing around 3,000 people at its Edge Lane site, before eventually closing in 2012 (Jaleel, 2012).

14 The experience of Martha's mother here is interesting, and typical of the social mores of the time (see e.g., Oakley, 1972); both of Martha's parents were employed by Plessey's – her father being a cable maker. However, her mother's role as a wirer was a more skilled role, and one which required staff to pass an aptitude test; as such, it paid the best wage of the two roles. In spite of this, as a woman, Martha's mother had sole responsibility for organising and paying for child care. She loved and was proud of her job, and wanted to keep it, but lack of support/engagement from Martha's father made juggling the journey and the full-time job impossible, particularly when she fell pregnant with Martha's sister.

15 A brown willow pattern plate procured from the rag man in the late 1950s is still in the family's possession; the plate was awarded to Pat, Maureen and Colette's brother-in-law Colin on the death of their mother, as he was the only family member who did not object to eating off 'the rag man's plate'.

16 Jackie feels fairly certain that her father experienced emotional difficulties, and quite possibly some mental health difficulties, and that barriers in the form of gender expectations and notions of masculinity prevented him from seeking help during the 1970s and early 1980s, when his alcohol and gambling issues were at their peak.

17 Peter refers here to the additional £10 paid on top of unemployment benefits to claimants who agreed to attend a year-long Employment Training programme (MacDonald and Marsh, 2005).

18 Martha has a theory that many of the teachers at Cantril High were 'rejects' from other schools; 'We had a professor who taught us physics and a doctor who taught us chemistry, and the music teacher had been from a public school, and the maths teacher…had his own formulae, and when I look back now I think that these were all people who were kicked out of schools that had really good reputations, or had breakdowns, or whatever, and they'd come along to this council estate as it was all they could get'.

19 The way that education has evolved in the Cantril Farm/Stockbridge Village area is interesting. When the new county boundaries came into effect in 1974, the majority of Cantril Farm was assigned to the newly created borough of Knowsley. Following the well-established and well-trodden path of subjecting the Liverpool area to social experimentation, Knowsley became a *de facto* test centre for New Labour's Building Schools for the Future venture in 2005 (Cobain, 2017). The worrisome conclusion that most of the children in Knowsley were kinaesthetic learners, i.e., they could not sit still, led to seven new secondary schools or 'learning centres' being built across the borough, with no classrooms – just wide spaces to facilitate big numbers of kinaesthetic learners. All seven establishments have now either been closed, or re-structured with traditional classrooms, as the experiment was deemed a failure after just four years (ibid.). Knowsley continues to have among the weakest educational attainment levels in the country (Weale, 2016).

20 Determined to continue with her education, Martha achieved an A grade for her A-level Law, and subsequently did another part-time A-level in Social and Economic History, achieving another A grade after talking six weeks' unpaid leave from work to study. This led to a law degree, for which she attained a first, followed by a PGCE and a long period teaching A-level Law. She was awarded her doctorate in 2018 and has since established an academic career at a North West university.

21 I interviewed Maureen and her sisters the day after Phil Scraton, academic, campaigner and author of *Hillsborough: The Truth* (2009) gave a speech at the University of

Liverpool to mark his honorary doctorate; Maureen and her husband had attended the event in honour of their son Gary.

22 Pens, in the sense of cages or impounds, were discussed several times throughout the interviews, mostly in connection with the docks and the 'pens' that casual dock labourers would be selected from if they were chosen for a day's work. Respondents frequently commented on the demeaning manner in which the dockers were treated, including ex-dock labourer John; 'It was like animals, I'll be honest with you, it was cages, and you had to put your book in, and you'd see them coming over, thousands of men, [it was] degrading'.

23 Throughout the course of my interviews, I have heard this derogatory phrase used several times, even though the area has been known by its new moniker, Stockbridge Village, since the early 1980s. Martha gave this example; 'getting a taxi home with my mum, and the first thing the taxi fella says is where are you going? Cantril Farm, oh you mean Cannibal Farm, you all eat each other there don't you, and that would really upset my mum, and that sticks with me'.

Bibliography

Balderstone L, Milne G and Mulhearn R (2014) 'Memory and place on the Liverpool waterfront in the mid-twentieth century' *Urban History* 41 (3) pp 478–496. doi:10.1017/S0963926813000734.

Belchem J and MacRaild D (2006) 'Cosmopolitan Liverpool' in Belchem J (ed) *Liverpool 800: Culture, Character and History*. Liverpool University Press: Liverpool, pp 311–391.

Brink S (1995) 'Home: the term and the concept from a linguistic and settlement-historical viewpoint' in Benjamin D and Stea D (eds) *The Home: Words, Interpretations, Meaning and Environments*. Avebury: Aldershot, pp 17–25.

Broomfield N (1971) *Who Cares?* Documentary. Available at http://www.liverpoolpictorial.co.uk/blog/liverpool-1971/. Accessed 30 March 2016.

Charleston S (2009) 'The English football ground as a representation of home' *Journal of Environmental Psychology* 29 pp 144–150. doi:10.1016/j.jenvp.2008.06.002.

Cobain I (2017) 'The making of an education catastrophe – Schools in Knowsley were dubbed "Wacky Warehouses"' *The Guardian*, 29 January 2017. Available at https://www.theguardian.com/education/2017/jan/29/knowsley-education-catastrophe-a-levels-merseyside. Accessed 26 March 2017.

Coolen H and Meesters J (2012) 'Editorial special issue: House, home and dwelling' *Journal of Housing and the Built Environment* 27 pp 1–10. doi:10.1007/s10901-011-9247-4.

Couch C (2003) *City of Change and Challenge: Urban Planning and Regeneration in Liverpool*. Ashgate: Aldershot. doi:10.4324/9781315197265.

Courtman S (2007) 'Culture is ordinary: The legacy of the Scottie Road and Liverpool 8 writers' in Jones D and Murphy M (eds) *Writing Liverpool: Essays and Interviews*. Liverpool University Press: Liverpool, pp 194–209. doi:10.5949/upo9781846314476.015.

Dorling D (2006) 'Prime suspect: Murder in Britain' *Prison Service Journal* 166 pp 3–10.

Douglas M (1991) 'The idea of a home: a kind of space' *Social Research* 58 (1) pp 287–307.

Dovey K (1985) 'Home and homelessness' in Altman I and Werner C (eds) *Home Environments*. Plenum: New York, NY, pp 33–64.

Dumbaugh E and Rae R (2010) 'Safe urban form: Revisiting the relationship between community design and traffic safety' *Journal of the American Planning Association* 75 (3) pp 309–329. doi:10.1080/01944360902950349.

Fazey C (1988) *The Evaluation of Liverpool Drug Dependency Clinic: The First Two Years 1985–1987*. Research Evaluation and Data Analysis: Liverpool.

Hanley L (2007) *Estates: An intimate history*. Granta Books: London.

Jaleel G (2012) 'Last employees at former Plessey Crypto site in Wavertree say farewell' *Liverpool Echo*, 2 April 2012. Accessed 17 May 2017.

Kearon T and Leach R (2000) 'Invasion of the "Body Snatchers": Burglary reconsidered' *Theoretical Criminology* 4 (4) pp 451–472. doi:10.1177%2F1362480600004004003.

Lee C and Stabin-Naismith B (2001) 'The continuing value of a planned community: Radburn in the evolution of suburban development' *Journal of Urban Design* 6 (2) pp 151–184. doi:10.1080/13574800120057827.

Lewicka M (2005) 'Ways to make people active: The role of place attachment, cultural capital, and neighbourhood ties' *Journal of Environmental Psychology* 25 pp 381–395. doi:10.1016/j.jenvp.2005.10.004.

Lewicka M (2010) 'What makes a neighbourhood different from home and city? Effects of place scale on place attachment' *Journal of Environmental Psychology* 30 pp 35–51. doi:10.1016/j.jenvp.2009.05.004.

Lund B (2011) *Understanding Housing Policy*, 2nd ed. Policy Press: Bristol.

Mac an Ghaill M and Haywood C (2011) '"Nothing to write home about": Troubling concepts of home, racialization and self in theories of Irish male (e)migration' *Cultural Sociology* 5 (3) pp 385–402. doi:10.1177%2F1749975510378196.

MacDonald R and Marsh J (2005) *Disconnected Youth? Growing Up in Britain's Poor Neighbourhoods*. Palgrave: Basingstoke.

Mallett S (2004) 'Understanding home: A critical review of the literature' *Sociological Review* 52 (1) pp 62–89. doi:10.1111/j.1467-954X.2004.00442.x.

McBane J (2008) *The Rebirth of Liverpool: The Eldonian Way*. Liverpool University Press: Liverpool.

Muchnick D (1970) *Urban Renewal in Liverpool*. The Social Administration Research Trust: Birkenhead.

Murden J (2006) 'City of change and challenge: Liverpool since 1945' in Belchem J (ed) *Liverpool 800: Culture, Character and History*. Liverpool University Press: Liverpool, pp 393–485.

Oakley A (1972) *Sex, Gender and Society*. Ashgate: London.

Pearson G (1991) 'Drug-control policies in Britain' *Crime and Justice* 14 pp 167–227.

Porteous D and Smith S (2001) *Domicide: The Global Destruction of Home*. McGill: Montreal.

Ravetz A (2001) *Council Housing and Culture: The History of a Social Experiment*. Routledge: London. doi:10.4324/9780203451601.

Rogers K (2012) *Lost Tribe: The People's Memories*. Trinity Mirror Media: Croydon.

Scraton P (2009) *Hillsborough: The Truth*, 20th anniversary ed. Mainstream Publishing: Edinburgh.

Thornton P (2003) *Casuals: Football, Fighting and Fashion: The Story of a Terrace Cult*. Milo Books: Preston.

Topping P and Smith G (1977) *Government against Poverty? Liverpool Community Development Project 1970–75*. Social Evaluation Unit: Oxford.

Tuan Y (1977) *Space and Place: The Perspective of Experience*. University of Minnesota Press: Minneapolis, MN.

Waghorn K (2009) 'Home invasion' *Home Cultures* 6 (3) pp 261–286. doi:10.2752/174063109x12462745321507.

Weale S (2016) 'Academy sixth-form closure to end A' level provision in UK borough' *The Guardian*, 29 March 2016. Available at https://www.theguardian.com/education/2016/mar/29/academy-shut-sixth-form-a-level-provision-uk-borough-knowsley. Accessed 26 March 2017.

Withers C (2009) 'Place and the "spatial turn" in geography and in history' *Journal of the History of Ideas* 70 (4) pp 637–658. doi:10.1353/jhi.0.0054.

Womersley J (1954) 'Some housing experiments on Radburn principles' *The Town Planning Review* 25 (3) pp 182–194.

7 The Liverpool home in the 'backward-facing' era

This second analysis chapter considers how experiences and understandings of home have shifted since the early 1980s. As outlined and conceptualised in Chapter Four, 1981 marks a turning point in terms of regeneration in Liverpool as, in the wake of high unemployment and social unrest, heritage and history are established as the new direction for regeneration. Thus, data and analysis presented in this chapter applies to experiences and memories of regeneration and home in the more recent past, and is again organised thematically, starting with the demise of futuristic optimism and the emergence of the past as the key to prosperity.

Taking stock and evaluating what we already had

It is far too simplistic to suggest that the end of the 1970s in Liverpool marked a clean break with 'forward-facing' policy and planning and the start of the 1980s heralded a clean slate; a new beginning of 'backward-facing' heritagisation and plundering of the past in the pursuit of prosperity. Rather, we can identify an overlapping period whereby the last stages of post-war optimism die out alongside an emerging notion that future prosperity lay in the historic buildings and cultural traditions of a wrongly maligned city and populace, thus far overlooked and ripe for redevelopment, regeneration, and fresh exploitation. However, it is helpful to identify the events of the summer of 1981 and their immediate aftermath as something of a pivotal moment in this transitional period, a crisis point which required immediate and significant action.

Respondents recognised that this action arrived in the form of Michael Heseltine, and expressed both relief and respect at his intervention:

Brian: Heseltine, because when he came around this estate apparently he was amazed, where are the pavements, and I will give him credit there, but, you can't take it away from the man.

Lynne: We liked him … it was Michael Heseltine came round about the riots,[1] and he was on the bus going round, and he said why are there people walking on the road, there's no paths, and people had been after them for years, guaranteed within the next few years they were all built.

Clare: Really.
Lynne: I mean I'm not a Conservative by any means.
Clare: No, me neither, but you've got to hand it to him, haven't you?
Lynne: Oh yeah, you can't take it away from him.

Martha has similar feelings:

Martha: I remember Michael Heseltine coming to the estate, because we had riots … me and Dad were home one night, and we could just hear all this noise and we went out, and it was a full blown riot; people with petrol bombs, throwing them, and then the police arrived on the number 11 bus, and I remember being just so shocked because I knew things were bad, but I didn't realise that people were so disaffected … but Michael Heseltine got wind of it apparently … and I remember him coming to the estate, I actually remember seeing him.
Clare: What year would that be?
Martha: I think I was 11, 1979,[2] when Thatcher very first got in.
Clare: And then Heseltine pitches up on his white horse.
Martha: And he *was* on a white horse though, I mean I'm not a Conservative voter, but I will say Michael Heseltine was our saviour, because the money got invested then, and basically it became Stockbridge Village and everything changed.[3]

The recognition and the gratitude expressed to Heseltine and his intervention contrast sharply with views presented in the literature, which essentially write him off as a meddling make-over artist[4] (Tulloch, 2011; Frost and North, 2013). Further, respondents noted that Heseltine constituted an alternative to the 'managed decline' approach advocated by the likes of Keith Joseph and Geoffrey Howe under Thatcher, as indicated in this exchange:

Lynne: Because if it had been Maggie Thatcher, no matter what she did.
Brian: It [Liverpool] was just condemned by her wasn't it?
Lynne: She wanted us to get buried alive, her!

Perhaps the most useful way to consider this period is as one of audit – taking stock of what was left in Liverpool, and how it could best be exploited, developed, and regenerated to turn around the city's fortunes. Essentially, the 'auditors' – namely Heseltine, the Merseyside Integrated Development Operation (MIDO), the Merseyside Task Force, and the Merseyside Development Corporation (MDC) (Couch, 2003)[5] – established three distinct groups of 'assets' that should be both nurtured and harnessed to bring prosperity back to the city: The River Mersey; the city's historic architecture; and the people of Liverpool. Significantly, the majority of the respondents identified all three of these groupings as valuable resources in the city and, crucially, as evidence of the city's distinctiveness and superiority.

The River Mersey is the whole reason for the existence of the city and, as such, is very dear to several of the respondents. Irene, Susie, and Ronnie all spoke of having an affinity with the sea, and this fostering a love of the waterfront. Seafaring, the dockside and its associated industry, the Merchant Navy and the Royal Navy have touched the lives of almost all of the respondents in one way or another, and are reflected in the testimonies provided, for example:

> I just love the waterfront, yeah definitely … I went on the ferry when the three queens[6] were in … cost an arm and a leg, but still it was worth every penny, just to be on the river with the three queens, it was just something, well something you would never do again I don't think, because the ferry went out to meet them, and came in with them, it was just so nostalgic. (Irene)
>
> When my dad was skint and all that he'd still get us up, come on, out and all that, he'd have us walking, like walk down the Pier Head, we'd walk right the way down, so we'd still be out and being tired, going on [dad's] shoulders, and you'd be tired, but you'd still be out. (Jodie)
>
> Recently I've had more time on my own, and [if] I'm feeling a bit down, I'll go down to Otterspool, and I'll just walk along the front, and I'll just look at this lovely city and I'll, it just lifts you, you know, it's just a lovely place to be. (Mary)

Similarly, several respondents spoke of the beauty and splendour of the many historic buildings in the city, particularly those at the waterfront:

> The buildings, the famous buildings, the cathedrals, the Liver birds, the George's Hall, all those historic buildings are very much part of the culture. (Susie)
>
> Look at the cathedrals, the Philharmonic Hall, the art galleries, the museums, the universities, the Albert Dock, the bombed-out church,[7] Liverpool has a long history. (Enid)
>
> I think yeah, the architecture in Liverpool is one of the best I've ever seen, I mean, took the kids last week to the museum, the art gallery, the buildings round there, down by the Pier Head, the Three Graces, and just about everything is beautiful. (Teresa)

Respondents also spoke about the people of Liverpool as distinct from 'other' people and, specifically having unique, valuable characteristics that could not be found in quite the same way elsewhere. Common character traits mentioned included friendliness and a welcoming, open demeanour, humour, community spirit, care for others, a focus on family, and a strong sense of justice. Some of the respondents identified gender-specific understandings of Liverpool people; men as 'tougher' than those from other cities, who

> would have been the breadwinner, very much family oriented, go out and earn the money, and they do the work to be able to bring home and provide

> for their families, your sort of annual little holiday somewhere, family parties, go the pub ... liking a drink, liking the football, probably liking a bet, probably liking a smoke. (Mary)

Liverpool women, meanwhile, were described as particularly 'strong', both emotionally and physically, in their roles as family matriarch (Teresa), community activists (Ronnie, Susie, Martha), and resilient 'survivors' (Jackie, Jodie, Mary). Interestingly, Jodie described herself as different from both her mother and her older sister, in that she is more forthright and prepared to challenge authority:

> See my mum and all that are very much like keep the peace, don't say too much, or when a doctor tests that, it's right, when a teacher says that, it's right, whereas I'm not like that, and our [sister] goes to me, how can you talk to a teacher like that, and I go, because the teacher is not my interest, [my son] is.

The notion of the glamourous Liverpool woman came up several times, with specific reference to high heels, faux tans, stylish hair and makeup, and a distinctive fashion sense. Jodie spoke about her own approach to fashion and style:

> I don't go out often, so why not, why not get done up nice to look gorgeous? Like I mean one-offs, going out now like I always dress a bit different and [what] no-one else had on ... [on a night out] I said to our [son] the other week, I bet she's not Scouse, and he went how, and I went I can tell by her heels, I could just tell, I am dead thingy over my high heels going out, I am dead particular over my shoes going out.

Jill discussed the focus on Liverpool women's sense of style and fashion from a different perspective:

> There is a certain look to a lot of young Liverpool women, with the thick eyebrows, and the huge eyelashes, the makeup, the straight hair ... this pressure to look a certain way ... the upkeep must be a fortune.

Ideas around Liverpool people being essentially 'different' to others was a recurrent theme. Christine, who lived for several years in Winstanley, near Wigan, suggested that Wiganers are

> strange people, dead weird, they are the type of people you could be sitting chatting to them like I'm chatting to you, or talking on the street having a conversation one day, the next day they'd walk past you, weird ... in the end I said to [husband], I said look, I'm going to have to go back to Liverpool, I said I can't stand it up here.

Both Joey and Dave stressed that people from Liverpool are so distinctive that they identify themselves as "Scousers; not English" (Joey); "I have heard and feel that people relate to the city in a way that transcends their nationality. When asked where we come from we say Liverpool and not England" (Dave). Jodie meanwhile discussed the differences between the people of North Liverpool and South Liverpool:

> So this area [North Liverpool] suffers with drugs and alcohol, and also dyslexia I believe is quite high in the North … so you've got the all the likes of that this side of town, culturally wise this side of town, I think this side of town is more materialistic as well. People know your business, don't they, in the North end of town … but down there it's a little bit different. That's how it's engraved into us [in the North end], you know like your windows have got to be clean because it shows that you've got a nice house on the inside, down the South end they're not bothered about anything like that.

Gary developed this idea of difference even further by discussing differences at the level of district:

> The first house [I lived in] was Central Avenue which is about two minutes away from where we live now and it was right in the centre of Speke, which is a bit crazy really because there's always been an east end and a west end of Speke which has been a bit of a rivalry, because it's the old end versus the new end, and when you're in the middle you've got friends in both ends, it became really precarious when I was a teenager.

Challenging the common stereotype that people from Liverpool are, amongst other things, unintelligent, several respondents insisted that Liverpool natives are quick by nature and have excellent judgement. Tony travels around the country for work, and maintains that

> No matter where you go with people, there's no-one as sharp as what, maybe not the cleverest, but with situations and that Scouse people pick up on it faster than anybody else no matter where you are, no matter what room you're in they're always on it before most others.

Carl, who lived in Australia for several years, spoke of a network of highly successful ex-pat Liverpolitans who prosper in any situation:

> It's typical of Scousers, everywhere I've been in the world, even in Jersey and I've been to Australia, you find that the managers, or the top fellas in companies, tend to be Scousers … we don't take no shit you see, if something is wrong [we'd] rather sort it out there and then … I'd rather work with Scousers than anybody.

Dave describes a Liverpool culture of "tenacity in the face of no agreement"; meanwhile Ronnie agrees that Liverpool people are intelligent and quick witted, particularly in the face of power:

> [We're] very clever, I think we're very good at very low level politics, and making your friends wherever you need to, and getting the support that you need, and not falling out with people ... rule is something you find your way around, it's not something to be respected, a rule is how people try to dominate you.

Thus, the general consensus amongst the respondents was that people from Liverpool *are* different from people of other areas in some ways, even though, as Ronnie points out, "we're not as different as we think we are". This difference was perceived by respondents as inherently good – that Liverpool people have some shared characteristics that render them somehow better than other populations. This is perhaps best encapsulated in Mary's suggestion that individual failure to display inherently good Scouse qualities results in people being metaphorically cast out of the Liverpool community:

Clare: So, you know if I said to you about James Bulger, what would you think there about the sort of negativity surrounding, could that have happened anywhere, or could it only have happened in Liverpool?

Mary: Now you see it's funny, I don't associate Venables and that as being Scousers, in my head.

Clare: Go on, tell me about that.

Mary: I don't know! See I've just realised now as we were talking about it, I can see Jamie[8] Bulger as a little Scouse toddler but I don't see them two as Scousers, it's as if they've come from outside and done it, and I don't really know why, cos I know they were from Liverpool, from the area and all that type of thing but I think in my head I've detached them as being, well you're not worthy to be part of the Scouse community so you're just not, you're binned off, you're just a pair of weirdos, sick weirdos who have done terrible, terrible things.

Although the majority of respondents did view the people of Liverpool as not only *different*, but *better than* other groups of people, there was also some recognition that in many ways Liverpool people are 'ordinary', and any suggestions of extraordinariness are problematic, particularly where the media are involved. Teresa addressed this issue using the term 'professional Scouser':

> I think some, I sound like I'm stereotyping Scousers myself, but some are professional Scousers aren't they? I mean I can't speak anyway, as broad [Scouse accent] as I am, I'm not posh in any way, but there's some really drag the accent out, and I think if, say, other parts of the country come down and film they always pick the worst one, they couldn't pick the likes of you,

> or just ordinary Scousers, but they have to drag them out from somewhere, and I think oh for God's sake … I feel a bit like embarrassed for people when they're like that.

Susie, however, recognised that *some* Liverpool people are not seen as a reputational asset to the city; rather, they are seen as a problem, a detriment to the city and perceptions of its culture and prosperity

> Most of the jobs I've worked in have involved vulnerable people … so for example for many years I worked with street sex workers and they were kind of designed out if that makes sense, so the areas traditionally where they worked became built up … so what happened was we had general dispersal then, and it made them scatter across the city which is dangerous obviously if we can't find them … and most of them were homeless which is the other population, and the homeless population obviously with the things like the gold zone,[9] and banning people from the city centre, how can that be right? They've as much right to be in the city centre as we have.

In terms of the Liverpool populace, then, only some are viewed as an asset – those who are able to fulfil the requirements of a city with prospects, a city on the move, a city with heritage and a rich culture. Liverpool people who do not fit this profile, who are not viewed as a reputational asset, must be 'airbrushed' from the picture and disqualified as a resource within the city.

Thus, notwithstanding some specific deviations, the populace of Liverpool was viewed by the respondents as a valuable asset; a resource inherent to the city which could and should be harnessed with the goal of economic, social, cultural, and political regeneration, just like the natural asset of the river, and the historical asset of the built environment. It is to the notion of the city's history, and its value in terms of regeneration, that we will now turn.

Establishing a future based on the past, and a culture authorised by primary definers

Arguably, the reclamation of the Albert Dock, orchestrated by Michael Heseltine in his *de facto* role as 'Minister for Merseyside', marks the pivotal moment in Liverpool's history when the city's past was identified as the city's future. Recognition that the restoration of the historic dock – revolutionary when it was first built – could act as a contemporary tourist and leisure attraction, was enough to attract private finance to invest in the regeneration (Couch, 2003). The Albert Dock was thus the first of a range of developments in the city to draw on the industrial past to create a future in a post-industrial climate. Avery (2007) notes that there was a level of cynicism about the potential success of the project, but, by the time the first site was ready in 1984, its opening coincided with two complementary projects which enhanced its value: The Liverpool International Garden Festival, a garden theme park developed on reclaimed industrial land at the South docks (Theokas, 2004); and the Tall Ships Race – a maritime festival celebrating historical vintage wind-powered ships (Longman, 2016).

Alongside the heritage-driven nature of the Tall Ships event, the Garden Festival drew heavily on both the industrial and maritime past of the city, as well as the legacy of The Beatles, 20-plus years since their rise to fame (Roberts, 2012). The success of these related events arguably kick-started the prioritisation of the past in the continuing regeneration story in Liverpool, and solidified the nascent trend towards history, heritage, and commemoration as the established method and direction of regeneration in the city henceforth.

The Albert Dock was mentioned on several occasions during the course of the interviews, usually in a very positive, affectionate way; although Ronnie did state that "I don't go to the Albert Dock really much, I don't really know what it's for". For Teresa, the Albert Dock is a place of beauty within her home city in which she can take pride, particularly as it is emblematic of progress that has been made since the 1970s:

Teresa: The Albert Dock is my favourite, I just love it down there, and I think it's more friendly, it was always friendly, I'm wrong saying that, but I think it's just lovely, it's a lovely area.
Clare: Yeah, it is, isn't it?
Teresa: Yeah, because we watched, you've probably seen them all if you're interested in that, the *Boys from the Blackstuff*?
Clare: Oh yeah.
Teresa: George's last thing,[10] did you see the Albert Dock there?
Clare: Yeah.
Teresa: Isn't it amazing what they've done?
Clare: Yeah it is.
Teresa: I think it's fabulous.

However, none of the respondents brought up the 1984 Tall Ships Race, and only one, Irene, recalled the Garden Festival of the same year:

> We went all the time … oh yeah, it was really lovely … all the different gardens, and things that were going on, and everybody's so friendly, and it was so easy to get to, you just went down to the Adelphi, you got on the bus, you were taken straight there, and you could come straight back from there and that, we went a lot straight from school … take a picnic tea with us, I really loved it.

The Garden Festival in particular was very popular amongst people in Liverpool; it was visited by 3.8 million people during its five-month life span (Wright, 2014). Thus, it is somewhat puzzling that only one respondent recalled the event when asked to reflect on cultural regeneration initiatives in the city. Possible reasons for this could be the length of time that has passed, or the subsequent rapid decline of the site after the event was concluded.[11] However, from the discourse emerging from this project it can be argued that both the 1984 Tall Ships Race and the

Garden Festival have been eclipsed by the subsequent narrative of 'Liverpool culture' in the ECOC 2008 context, dominant since the accolade was announced in 2003. Heritage and culture, leisure and entertainment are predominantly viewed through this lens; filtered through an established discourse whereby what constitutes culture and what constitutes leisure in Liverpool is dictated via statecraft and governance by primary definers – Liverpool City Council, Peel Holdings, and other stakeholders seeking to profit from the harnessing of a Liverpool-specific heritage, culture, and leisure scene.

Starting with the appropriation of the Mathew Street Festival, which had emerged organically in 1993, into the statecraft-endorsed Liverpool International Music Festival in 2013, grass-roots culture appears to have been relegated to a non-cultural status, as 'real' culture is administered, facilitated, and validated by Liverpool City Council and its partners. Some of the respondents recognised that a certain level of appropriation and 'sanitisation' of Liverpool culture was apparent within various regeneration initiatives. Dave suggests that, while the grass-roots culture, the "creativity of the natives" that he felt part of when he lived in the city's Georgian quarter, is still in place, other cultural aspects of Liverpool have been spoiled:

> I feel, and I'm not certain that this is the case, that this regeneration is sourced from interested parties outside of Liverpool. I am very disappointed in the upgrade of the Everyman Theatre and I feel that this has excluded the local people. I know that I am not attracted to it anymore and yet I was a frequent visitor when it was a grass roots theatre that was community based. … I certainly don't feel that I am part of what's happened to the city, it's like it's been taken over by a new host.

Susie also misses the 'old' Everyman, and sees its re-invention as part of a wider process of commodification with regard to the organic, spontaneous culture inherent to the city. Specifically, Susie is aggrieved at the commodified commemoration of nightlife in Liverpool, whereby nightclubs like Cream, 051,[12] and The State[13] live on as 'events' rather than a physical entity, and a series of 'reunions' are created to invoke the night-time culture of the past:

> I loved The State … we had a reunion last year … but someone's commodified The State, having reunions and all that kind of stuff, I think we were charged about £40 each for a ticket, and it was never about all that, it was about people who liked Echo and the Bunnymen and that kind of scene. … I loved it, and then people come along now and, you know, getting ripped off.

In a similar vein, Peter reflects on the Pier Head stretch of the waterfront, and an existing night-time culture which has been extinguished:

> What springs to mind is the Pier Head area, the ferries still come in and all that, again it's added on the Albert Dock, and the Liverpool museum and

> what they've got down there now, but the actual Pier Head area and that, it used to be a place where you could go in the night, and it would still have an atmosphere, now like it hasn't got an atmosphere, no-one goes down there of a night. They used to have little local bacon butty places open in the night where people would go like taxi drivers, or even like teenagers, and it would have its own little atmosphere, all the buses would park there, but now they don't, it's just bare, and it's just a tourist walk.

There is, however, evidence of the 'grass roots culture' that Dave speaks of still apparent within the testimony of respondents, for example Pat, who frequently spends Sunday afternoons in a city centre pub, The Hole in the Wall:

> They used to have dances in the schools, and we'd go to dances and that in St Anthony's and St Sylvester's, and St Bridget's, I'd go round them all, but the lads used to have guitars, and they'd play in the street, now the lad that played guitar, I still see him on a Sunday, The Hole in the Wall, we have a great time in there, I go with my friend [name], I go to mass at half 11 then I dash up to get the half 12 train, and we go there, it's great, I love it, yeah, just a singalong with Kenny, he's played all the pubs, The Atlantic and all them, your dad would probably have known him, and he goes I just love Scotland Road because he was brought up down there as well.

Consequently, it can be seen that Liverpool's past is now established, in a variety of ways, as central to both its present and its future. Crucially though, only certain versions of, or elements within, the past, are validated as 'culture' and 'heritage'; worthy of commemoration and commercialisation. The Beatles were a bigger mainstream act and commercial success than Echo and the Bunnymen – thus, they are worthy of commemoration; the Everyman Theatre was successful, but small and intimate – thus, it must be re-developed for an audience of 400 and its old seats auctioned off as mementos (BBC, 2011); the Pier Head had a night-time 'scene' beloved of locals, but it was depriving business of making capital out of the views over the Mersey – thus, it must be re-modelled with hotels and restaurants. The past is constructed as vital to Liverpool's future prosperity, but some elements of the past are deemed either not worthy or not profit making; therefore, they are airbrushed from the version of the past prioritised and co-opted by "the regeneration professional" (Furmedge, 2008, p82). The resurgence of the old-fashioned, Scotland Road-style singalong as described by Pat, can be viewed as both resistance and a challenge against the imposition of a validated culture, refined for respectable and exploitable consumption.

Re-prioritisation of the city centre to the detriment of districts

When place is first established, it is established small – a valley, a woodland clearing or, in the case of Liverpool, a river. Liverpool as a place developed solely because of the existence of a river, with easy access to the Irish Sea and, beyond that, the Atlantic Ocean. As such, Liverpool started on the banks of the Mersey, and radiated outwards in a semi-circular pattern. In simple terms Liverpool started at the *centre*. Because of this historical pattern of development, it is unsurprising that the majority of ideas around Liverpool's past focus on the city centre, where place was first established. Thus, processes of regeneration which prioritise culture, heritage, and legacy will emphasise the city centre and its historic built environment. Arguably, the result of this is the neglect of districts and suburbs; places further out in the pattern of concentric semi circles. During the 'forward-facing' period, districts, some at least, were prioritised via the policy goal of modern housing for all, and the city centre, although not ignored, was only part of the master plan to prepare Liverpool for the future. 'Backward-facing' regeneration, however, has no real need to consider the districts, as 'the past' is conveniently encapsulated within a few square miles in the city centre. This section addresses the theme of prioritising the city centre over the districts, and what this means for respondents.

None of this is to say that the districts, from inner areas like Edge Hill, Dingle, Kirkdale, and Walton to overspill developments like Speke, Stockbridge Village, and Kirkby have been entirely ignored since the conceptual turn from 'forward' thinking to 'backward' thinking. Indeed, a variety of phenomena, from maintenance and repair of the built environment to organic economic revival, are apparent in the districts from the early 1980s onwards. A very early example of beneficial change came as a result of Michael Heseltine's post-disturbance visit to Cantril Farm:

> There was real investment, they turned all the houses around, so my mum's house [you] used to drive in, and the same thing, you'd have to go round the back, but now the houses were all turned round to face the road. (Martha)
>
> At the time, this is complicated, at the time [when we moved in] the kitchen was at the front, but since they improved all the security measures to try and improve the reputation and everything, and rectify the faults the designers built into the buildings, they actually turned them around, so the back is now the front. (Roy)

Thus, alongside the rebranding from Cantril Farm to Stockbridge Village,[14] the houses were re-modelled, and pavements were created to provide the area with a more traditional road layout, with pavements leading to front doors rather than the Radburn-inspired narrow, often covered, walkways.

Modern houses were dismissed by many of the respondents as lacking in quality and authenticity; Jill, for example, referred to the 1980s-style housing standing on the site of her first home in Pecksniff Street as "little red boxes", and Irene used the term "matchbox houses". Many such houses were built in the 1980s under the *de facto* Militant Labour council, in the inner areas on the outskirts of the city centre and, at the request of those who would go on to live in them, in the *Brookside*[15] style; detached or semi-detached, with a front and back garden, on a curved cul-de-sac facing away from the main road. While such housing estates are derided in some quarters now,[16] they continue to be popular with those who live in them, particularly those rehoused in them after living in poorly maintained tenements. Peter moved with his wife and three small children from a ground floor flat in Vauxhall Gardens to a three-bedroomed semi-detached 'Hatton' house in Benton Close, Kirkdale, in December 1986:

Clare: Was it good news then when you went into Benton Close?

Peter: Well it was unbelievable news, I always thank the local council at the time, personally, this council get called a lot, the so-called Militant council of the mid 80s.

Clare: They built a lot of houses though.

Peter: Well I'm eternally grateful for Derek Hatton and all the other 48 councillors, because in fact actually I think they're the only people who took Thatcher on and beat her … and they did build all these houses, and get rid of all these rat-infested tenements, which they were at that time, they were all decaying, and there were rats in people's back kitchens running round. … I'm eternally grateful because the amount of houses they built around the inner city.

Clare: About 5,000.

Peter: Oh God, it must have been, yeah, and I think they'll go down in history as like the best local council ever, and I know other cities followed, the likes of Sheffield and Manchester with the inner city building of housing, semi-detached and closes.

Clare: Yeah, you're right, they promised they'd build those houses, and they did build those houses.

Peter: They said look, these people need houses, we told them, that's what we got elected for in our local manifesto, and we're doing it, and thank God he did, and it was a relief for lots of people … the happiness that gave people at the time to get out of what they lived in, into them, with back gardens with grass, and they could sit out, oh my God, that's something we dreamed of.

Thus, it is important to note that some elements of the conceptual 'forward-facing' era of regeneration continued into the 1980s, the Militant-built closes and cul-de-sacs being a clear example. They are beyond value to people like Peter, even though they are considered "non-productive" (Heseltine and Leahy, 2011, p38), therefore a waste of capitalist opportunity, to others.

Although many districts plummeted into further decline during the 1980s and 1990s, by the turn of the century several districts started to experience a level of nascent revitalisation. Jodie worked for several years in the Lodge Lane area of inner Liverpool, first as an immigration advisor in a law firm, and subsequently in several third sector roles. She notes some very positive changes in the area since austerity measures were imposed by the Conservative/Liberal Democrat coalition government from 2010 onwards, compared to when she first worked there:

> When I first went down there you had the shop, you had a few pubs that were derelict because obviously that type of area is not going to make money,[17]

> and you had one kebab house that everyone used to go to in the south end ... now you've got the barbers, the credit union, you've got smart restaurants down there, the ice cream parlour ... it's business now, and I think everyone is wanting to be part of that business now, so they're not looking at it as if it's an easy thing, you know, you want to make money, let's make a business, and if *they* can do it, *we* can do it.

Significantly, Jodie attributes this turnaround to funding cuts resultant from austerity policy at Westminster level, and a subsequent battle for funding among third sector organisations:

> That area has had a lot of money pumped into it, and things don't change, it started to become the norm to be given things, rather than earning it ... so I think when money became tight it was a sure shock, well I've got to get up and work for this now, I've got to be able to do this, I can't just go to the community (centre) and maybe have a dinner that's less than £2 and stuff like that, I've got to work, and to be fair I think that's probably done it more justice because look at Lodge Lane now, people have gone into businesses.

From Jodie's perspective, then, the renaissance that Lodge Lane is currently experiencing is an organic development resulting from inherent entrepreneurialism, awoken by austerity after lying dormant as various third sector groups competed for funding to 'save' the area from the top down.

Other districts of the city have had, since the turn of the century, extensive and intensive investment and tailor-made regeneration attention, often resultant in very little real change. Mary, Rita, and Irene all spoke about Kensington New Deal (KND), a regeneration initiative developed to rejuvenate the Kensington district, an area blighted by long-term unemployment, transience, poor health, crime, and dereliction (Allen, 2008). The New Deal initiative was established in 2000 as part of the Blair government's New Deal for Communities (NDC) programme and was designed to tackle deprivation in all its guises within the L7 area (Couch, 2003). Although the perceived need for the KND was multi-faceted, Rita remembers the start of the initiative as being triggered by the conduct and activity of one particularly notorious family living in the area:

Rita: Well, how it first came about, that all started because we had, we didn't have a very nice family down the street, and they used to cause, they caused *havoc* in the street didn't they Mary, the ******s.[18]

Clare: I can remember the ******s! Yeah, there were a lot of problems, weren't there?

Mary: Their house has gone now, it got knocked down didn't it? Yeah, it's now a bit of grass.

Rita: Yeah, opposite the school, and they were lovely houses, they were lovely, there was about 8 houses or 10 wasn't there Mary, at the top of Brae Street, and they were in one of them, and they wrecked theirs and they wrecked next door, and they're [pulled] down now, at the back of our [daughter's].

Clare: How many of the ******s were there?

Rita: Well there were 3 lads wasn't there, and they caused havoc but then they had, it was their friends that brought, and they had family through the park that came ... they were scallies, they were *all* scallies.

Irene, however, indicates that Kensington was selected as an NDC as a result of its demographics from the perspective of the government, and that Rita has conflated two issues together:

> I wouldn't have said that it was because of [the family in question], it was just regeneration of the area ... we were never told why, we were just told we were chosen, and that was it.

From the perspectives of Rita, Irene, and Mary, all Kensington residents of long standing, the KND was, contrary to the conclusions presented on their own website,[19] characterised by limited success, wasted money, and broken promises. Hopes had started out high, but were ultimately dashed:

> I think there were promises, there were promises made at the beginning, particularly in respect of housing, and there were radical plans, it was all going to be this, and it was all going to be marvellous, it was all going to be wonderful, and it was going to be like this sort of – oh I can't even think of the word, sort of this serene, ideological place that we were all going to live in, and I remember going to some of the planning meetings, and there were artist's representations, this is what your street is going to look like, and it all went to shite, and it was really, when you think about it, they were completely unrealistic ideas and plans in the first place, and again I think the expectation was like right, it's going to lift the neighbourhood and it's going to lift this and lift that and, you know, there's only so much that it can do, there's limitations, I think it tried to be all things to all people, but it couldn't fulfil, and so people felt disappointed by some of it. (Mary)

'Radical' plans that did not come to fruition included extensive building of new 'family' houses, leading to significantly higher numbers of children attending the local primary school (Mary); refurbishment of all the properties on Holt Road (Irene), and the modernisation of houses in designated streets, including Mary's:

> Even the modernisation of ours though, it's all fur coat and no knickers, it was to make the fronts look nice, so it looks nice for people driving past and driving through, and the quality of what they actually did, the quality of the windows that they put in was *less* than the windows they took out, because

> our windows were like over 10 years old, and they were out of warranty and all that, we were like "oh we may as well" and the windows that they put in were shite, the door isn't wonderful, fur coat and no knickers, and we all said at the time that it was fur coat, and everyone was fully aware that it was, even down to the fact that you had, you were asked what colour paint you wanted on the front of the house, but you were sort of told that everyone else in the street was having this colour, and it was like, I remember in Jubilee Drive someone wanted a different colour, and there was murder! Because they wanted it to all look uniform and pretty, rather than actually like this person wants.

Irene cites the wholesale replacement of windows in certain streets as an example of money wasting and financial stupidity;

> They got the windows free, if you owned your property you got your front windows free, but the thing was, if you sold your property within seven years, you had to pay a proportion of the money back, that is not standing now obviously, because it's over the time,[20] but it didn't stand anyway, because the office that was dealing with it folded, so there was no way of them claiming the money back, so people were selling their houses, but weren't paying the money back, so it was just a farce really.

Patterns of house clearing before money was in place to build replacement properties, that first emerged in Liverpool in the 1950s and 1960s, resurfaced during the KND, and compulsory purchase orders were once again issued, causing the same amount of stress, emotional upset, and financial loss as they had during the 'forward-facing' era. Mary spoke about the experiences of her colleague:

Mary: When it was first going, "well, knock all this down", and people were moved out including one of my friends at work lived there, and she then had to get a mortgage at the age of like 55, because the house that they wanted to buy in the [Old] Swan, which was a comparable size house, comparable area really, but it was more money than they were getting off the compulsory purchase order, so they had to get a short mortgage to be able to move up there, their house was one, they were forced out, and most of the people in her street were sort of the same.

Clare: What street was that?

Mary: That was Nuttall Street, yeah, and they were all sort of forced out, and there wasn't really …

Clare: Did she not really want to go?

Mary: She didn't want to go at all, she was one of the last ones, she held on till the end, there was only her and another person in the whole street, all boarded up, which then had an impact on the money that she got, because the market value had dropped.

Nuttall Street, and the area immediately surrounding it, was an area identified by Mary, Rita, and Irene as a part of Kensington which should have been renovated rather than demolished:

> They've all come down and new houses have been built, and that was a bit of a shame because they were lovely houses, they just needed a little bit of attention, and I think the houses that they put up, they're nice enough and they've got gardens and it's what people want this that and the other but that's a bit of a shame, they were good quality houses because I knew people who lived there, and they weren't like some of the ones up on the front, like the flats we used to live in and all that type of thing, they *were* run down, they were damp, there were problems with you know, structurally and all that type of thing, and it would have cost more money to get them fixed than it would to bring them down, whereas the Nuttall Street ones, the majority of them needed just a bit of TLC, and they would have been, so it's a shame that some of them have gone. (Mary)

So, just like Everton, Dingle, and Kirkdale before it, 50 years on, swathes of Kensington have been bulldozed without real need, and against the wishes of many residents.

While Mary and Irene agree that the new medical centre and the new school, both built under the KND, are a welcome addition to the area, and a valuable asset, they disagree significantly on the question of the widening of Edge Lane. The main arterial route into the city centre, bringing traffic directly from the M62 (Couch, 2003), Edge Lane was widened from a single carriageway to a dual carriageway under the KND in an attempt to alleviate traffic congestion. Despite having reservations at first, Mary likes the result:

> I like the changes to Edge Lane, particularly my little bit, I like the fact that it's opened up, it's got a nice, it's got a better vibe about it, it's got a better feel to it, it doesn't feel as deprived, although it still is, it doesn't feel, I think it just doesn't look as rough, it just doesn't look as run down, it doesn't feel, you're sort of walking round and you're not, you've not got as many sort of dank, sort of abandoned streets.

Irene however sees the widening of Edge Lane as a folly, and a waste of money:

> Of course, Edge Lane was a waste of money, why they pulled all those beautiful houses down on Edge Lane to grass it is just something else, I mean that was just totally ridiculous .... Edge Lane is no different to what it was twenty years ago, it's still a car park ... no better off at all.[21]

For Irene, much of the KND was event driven, rather than driven by substantial change:

> I feel that there was a lot of money wasted, they did a lot of events like choosing the best person in the area, and you had to nominate people who are a

> good neighbour and all this thing, and they had these big events at the hotels in town, and to me it was just like that was money wasted, that could have been spent in better places.

The trend towards event-driven regeneration is clearly identifiable as a central component of Liverpool's backward-facing regeneration, and will be considered later on in this chapter.

One of the factors considered in selecting Kensington as an NDC area, and a factor acknowledged by Mary, Rita, and Irene, was the reputation of the district, both in terms of its local and its national reputation. The image and standing of districts within Liverpool is something that all the respondents displayed an awareness of in one way or another. Jeni spoke about the nicknames her parents have for each other, one being from Bootle and the other from Everton:

Clare: So were your mum and dad from Bootle?
Jeni: My mum was, my dad was actually a townie as my mum calls him, what did he used to call her?
Tony: A buck
Jeni: Bootle buck, and she used to call him a townie, because he was from Netherfield Road.

The terms 'buck' and 'townie' are both colloquial, derogatory, sometimes affectionate terms; the former commonly used in relation to people from overspill developments (for example, Huyton buck; Kirkby buck), the latter reserved for people from the inner areas in Liverpool, particularly the densely populated terraced streets and tenements. Respondents displayed commonly held views about which districts are 'nice' and which are 'rough', for example, Christine talked through her decision-making process when deciding on where to buy a house in the Huyton area:

> I didn't want to go the other side of Huyton because I'd seen like different areas, and round here, in between here and Broad Green, it was always nice, Pilch Lane was always nice with the trees … like St John's estate, I don't think that had a very good name, and I think Bakers Green, the other side of the Bluebell [pub], I think that had a bit of a name as well, so I knew I wanted to get away from that really.

In a similar vein, Martha discussed the differing reputations attributed to different areas, and how she responds to them:

> So my son's getting to know this girl, and I said, oh where's she from, is she from Rainford, but she's from Kirkby, and I thought to myself oh no, not from Kirkby, isn't that awful! And I stopped myself and thought, what are you doing … we're odd, Scousers, aren't we, sometimes, I take great pride in the fact that I'm from Liverpool, and I'm proud of my accent, but I don't particularly want to live in parts of it, if that makes sense.

Thus, there was awareness amongst the respondents that some districts were flourishing whilst others struggled. Some respondents however identified examples of decline which seemed to be more widespread, including local shopping areas which, at one time, were earmarked for expansion and development (Wilson and Womersley, 1977; Couch, 2003). Empty shops, and the growth in particular types of businesses in local shopping areas, was raised by several respondents, including Rita and Irene, who both lamented the decline of Kensington high street:

> I think a lot of it changed when they got all these outlets like Speke,[22] and places like that, the small shops couldn't cope then, the supermarkets came then didn't they, and shops couldn't cope then, so it was a shame because you never went to town, Wavertree Road was the same, Holt Road was the same, we never had to go to town for anything … it's just disgusting now. (Irene)
>
> They're all foreign places now, all foreign shops, there's all kinds of shops isn't there, hairdressers, nail places, phone place, and they're all foreign, Polish, Polish shops and that. (Rita)

Both Susie and Ronnie expressed similar views about Walton Vale, to the north of the city centre, which has visibly declined in recent years. Susie acknowledged this; "I love Walton Vale, I did love it more though when it was better, yeah, it was great, but it's dying now". Ronnie recalled that "Walton Vale was where we'd go if we didn't have time to go to town", but now says of the area:

> Oh it's terrible isn't it, like there's a secret plan going on to rubbish it, everything looks like it's dying a death, it looks like it's all been bought by some mystery land owner who's planning to turn it into a Sainsbury's or something … it's being deliberately ravaged, deliberately blighted.

The decline and closure of local pubs was also discussed by respondents. This is a nationwide phenomenon, attributable to a range of factors including the legislative change regarding smoking in public places and changing trends in terms of leisure pursuits and entertaining at home (Andrews and Turner, 2011). Mary suggests that the pub decline that has occurred in recent years has impacted on the nature and quality of socialising at the local level in the Kensington area:

> If you look at Kenny [Kensington high street], you used to have six, seven different pubs, where it would be like party til whenever, and it would be people of all ages, barely legal to barely alive, and everyone in between, and everyone would be having a damn good hooley, and it would just be like that, whereas now there's like two pubs, well three pubs, when there was probably like about nine, so like the opportunities for it have got smaller and the venues have got more limited.

Jodie, Gary, Peter, and Tony, however, attribute a wider range of social phenomena to the decline in, and disappearance of, local pubs. Gary notes that money spent in local pubs was "money spent in the community". For Jodie, the local pub trade dying down has led to community fragmentation as, for example, local and children's football teams can no longer rely on sponsorship from the local pub for, for example, strips and kit. Tony also spoke about detriment to communities:

> You don't even go for a pint anymore, no one goes for a pint, so no one mixes, you hardly know your neighbours because you don't go to the local pub, and you don't go to the local pub because it's loads of money, and others don't go because they can't smoke or whatever it may be, it's cheaper to sit in the house or whatever, so you don't mix, you don't have the community spirit as much as you had years ago, because you don't know everyone.

Peter developed this line of thinking further:

> I've always been quite cynical about the amount of pubs that have shut down, and not just pubs, but like community centres and churches, and anywhere where the people gather, and to me, the way I think, as I say, I can be quite cynical at times, I sometimes feel it's like to stop people meeting, because that's where people express opinions.

The sentiment amongst these respondents is clear: The decline in the local pub trade throughout the districts has had a detrimental effect on communities, and, for Peter at least, is part of a deliberate drive to reduce community cohesion.

On the subject of socialising, Peter noted that many pubs in the city centre continued to flourish, and again took a cynical viewpoint:

> I think the inner city, I think the areas are missing the old-fashioned pubs that have all been knocked down, and that togetherness and that I mentioned earlier on, but town itself again, if I look at it cynically, it's like the hierarchy, and the powers, they want us all to be in one place at one time, all supervised with cameras everywhere.

Here Peter has hit on and encapsulated a clearly identifiable facet of contemporary Liverpool regeneration – people coming together must do so in the city centre, to ensure that (a) the city centre remains the focal point of the city, both for its inhabitants and for the wider world, and (b) both the surveillance of the city's populace, and the inscription on Liverpool people of the contemporary approved and validated version of Liverpool culture, is facilitated. Prioritisation of the city centre leads to ignorance regarding life in the districts, which then perpetuates the drive during the 'backwards-looking' era towards identifying as Liverpolitan at the *city* level as opposed to the *district* level. It is this preoccupation with, and emphasis on, the city centre during the 'backward-looking' era to which we will now turn.

Liverpool city centre is viewed almost universally by the respondents as characterised by beauty, particularly in relation to the waterfront and the historic architecture that dominates it. Some discussed the city's use as a film set, including Rita, who was aware that parts of the city centre are often used to portray New York, noting that "you feel good that it's, you know, Liverpool on the telly". Dave, on the other hand suggests that some of the city's authenticity is lost due to such developments; "The word I was looking for to describe the city nowadays is a 'film studio lot', a façade that hides its history". The 'beauty' of the city centre, together with the "café culture" (Susie) that has emerged during the 'backward-looking' period, is an incentive to live in or close to the city centre for some, including Susie:

> I want to move to the city centre, I want a flat … down by Brunswick [Dock], I like it down there, I like some of the apartments down there, because the water just draws me … . I work in the city centre, and I'm getting older now, so I don't want to have to be travelling where I don't have to, I'll go out, most of my social life centres around here because I like going to Bold Street to eat, and then you've got FACT and all the other things, the local theatres and stuff.

Ronnie noted that the large, multi-storey, and ornate buildings on Church Street, whilst having large shops on the lower floors, are empty above, and would make good homes:

> I wouldn't be above Coming Home[23] turning them into hundreds of homes, then you've got people who live in the place going down to have their tea in the cafes, you've got people who can live in the place, concerned about what happens, then [the area is] overlooked, and then it becomes, it's much more of a European way of living, and it just feels proper.

Paul posited that although many people would like to live in the city centre, particularly as he perceives the city to feel 'safer' in more recent years, most would not be able to afford to. Ronnie notes that this is particularly the case at the waterfront where, as time has gone by, unforeseen difficulties have emerged:

> People who've been living at the Kings Dock, the Albert Dock and the rest of it since they were yuppies in the 80s; they're now getting quite elderly, but the whole waterfront, there are no doctors, there are no playgrounds for your children, there are no parks, there's hardly any public space actually, so other than the view of the river … it hasn't got a sense of place, it hasn't really, they're like a collection of more or less gated communities.

It would seem then that the city centre is not yet geared up to be 'home' in the sense of a district, in that it does not have the required amenities, particularly with regard to children, the elderly, and the infirm. The districts are still *needed*, as they are better able to fulfil the needs of a wider community in terms of health care, education, and so on, but they are not fully *acknowledged* as such, as 'home' discourses in the Liverpool setting continue to be dominated by the iconography of the city centre.

Other than visits to either Goodison or Anfield football stadia, a 'Magical Mystery Tour' to the childhood homes of The Beatles, or perhaps a visit to Lark Lane[24] and Sefton Park, tourists as well as residents are also firmly encouraged to focus their attention on the waterfront and the city centre (Furmedge, 2008). The tourism in the city that has developed steadily since the early 1980s was acknowledged by the respondents and was, in the main, welcome, both as a source of income and a source of pride:

> There's a lot, a lot of restaurants, a lot of hotels, even just down to like, especially if you go down to the waterfront, you can see, feel the push towards tourism, you know, the tuk tuks and the open top bus, and we used to have the duck bus,[25] and you know all those sorts of things, they're all aimed at getting people to spend, and you know the Beatles Story and all that type of thing. (Mary)
>
> One thing that I've noticed is people, and you can tell they're on couples, so you might say to Jeni and Tony right we'll have the girls, you go and have a weekend away or whatever, like go into the city, enjoy the city, you see that a lot, and I like that. (Teresa)
>
> I think it's fab, I think it's good, I would never have thought that we'd take off as a tourist venue, but I like it, I like the fact that you can go to the cathedrals, and into town, and hear all kinds of languages … . I went down to the Pier Head the other week, and there was a ship with 3,500 passengers there [docked], but it does no harm to have those 3,500 people come and spend their money in the shops in Liverpool. (Ronnie)

Again, though, none of the respondents discussed tourism beyond the city centre; the focus was very much on the hotels, restaurants, bars, and shops in the city centre, together with the waterfront and associated historic buildings. In framing the city centre as the hub of tourism in the Liverpool city region, Liverpool City Council and its partners have followed a pattern established by other waterfront cities that have undergone a conscious, planned regeneration, for example, Glasgow (Doucet, 2013; Varna, 2014), Belfast (Coyles, 2013), Dublin (Byrne, 2016), Barcelona (Degen and Garcia, 2012), Cape Town (Ferreira and Visser, 2007), and Sydney (Prior and Blessi, 2012). What each of these waterfront cities have in common is that they are post-industrial and have all reclaimed a decaying industrial waterfront and reinvented it as a "playground" (Ferreira and Visser, 2007, p227). Joey, who loves and is very proud of Liverpool's waterfront reinvention, makes an interesting point in his testimony that "to regenerate an area where there is nothing to begin with is just *development*; it is culture and history that make it *regeneration*" (emphasis added) – however, Liverpool's city centre, and particularly its central waterfront, have arguably undergone extensive changes which go far beyond regeneration and well into the territory of development.

Development of the city centre, as opposed to regeneration of existing structures, was highlighted by many of the respondents. Several referenced the plethora of newly built hotels, particularly towards the waterfront, and some expressed concern that they were unsustainable – "but what happens if the bubble bursts when all these hotels and things, seems to me to be too many" (Lynne). Another form of development noted by respondents is student accommodation, usually multi-storey and capable of housing huge numbers of students, their proximity to the city centre making them attractive to students in terms of transport links and easy access to both leisure facilities and potential sources of part-time work. Carl approves of the growth in purpose-built student accommodation, noting that owners "must get five grand a year off each student, which is very good when you think about it". In the main, though, respondents expressed concern at the proliferation of dedicated student housing that appears to be 'swamping' the city centre, as encapsulated by Susie:

> I think the other thing is student accommodation, we go mad don't we, because that's all they're building, I passed a building the other day with [husband], can't remember where it was now, I was like, that's closed up, getting redone, and I went, I wonder what it's going to be, and [husband] went, here you are, there it is, student accommodation, so it's just constant, but it's in the city [centre], so again it's like that space is being taken out of use … you know, Cream and everything, Cream means a lot to this city, and that's gone now because they're building more student accommodation on it, they're literally everywhere you go.

However, the city centre development that most respondents identified was, unsurprisingly, Liverpool One. Lynne and Brian described their one and only visit to Liverpool One, a few years ago while they were waiting for their car to be serviced:

Brian: Gets off [the bus] at Liverpool One, and we're going round Liverpool One, do you know what Clare, if any security were watching us.
Lynne: On them CCTVs.
Brian: They'd be saying …
Lynne: They'd be hysterical.
Brian: We were up escalators, down escalators.
Lynne: Same one!
Brian: And then we got on the one to the Odeon Cinema which is massive.
Clare: Yeah, the big one.
Brian: And we got up there, and I think that's the only thing up there is the cinema, so when we got up there.
Lynne: Back down! And we ended up going up and down that one escalator because they all look the same to us!
Brian: Anyway, we got in the bus station, how can you get on the buses?
Lynne: Everything was closed up!

Brian: There were no gaps! Next thing the bus comes in.
Lynne: And the doors open!
Brian: Honest to god, you wouldn't believe.
Lynne: We were just like tourists in our own city.

Lynne and Brian's amused bafflement at being faced with negotiating that area of the city centre is indicative of the vast changes to the built environment which have rendered it unrecognisable from its previous form, although Ronnie acknowledges that

> it has actually reintroduced a Liverpool street pattern we lost in the Second World War, because basically all of Liverpool One was still a bombsite till it was built, so it has reintroduced the street pattern, it's linked the riverfront to the city.

However, respondents who discussed the development displayed views ranging from lukewarm to downright negative:

> I don't like Liverpool One right away, no I don't like that, I don't like the way it's set out, they should have gone to Milton Keynes and seen the way Milton Keynes was done … . I can't understand, there will be a shop here and you want to go there, and you've got to go all the way down and round to get to it. (Irene)
>
> Where all Liverpool One is, I liked Chavasse Park[26] and all that space, I loved it … . I hate Liverpool One, hate it. (Susie)
>
> I can't stand Liverpool One, I can't stand it with a vengeance, I don't like our town centre now the way it's gone, I think it's completely lost … . I just don't like Liverpool One, I don't feel Liverpool is unique as much anymore. (Jodie)

Ronnie has been able to draw some positives from the Liverpool One development, even though he never shops there:

> The best thing about Liverpool One is it's created space in Bold Street for loads of interesting independent business to open, so I actually never go to Liverpool One, I can't tell you anything that I would want to buy in any of the shops … . I'll often say to [wife] I'm going to town, I only mean I'm going to Bold Street, that's it, that's town, so I think that's been good, and I think that the city has to be everyone's city, so it has to be St John's Market if that's what you want, it has to be sipping coffee in Bold Street if that's what you want, it has to be going on a bling binge with your mates in Liverpool One if that's what you want.

Thus, the Liverpool One development has freed up space and buildings in other parts of the city centre for other, perhaps less corporatized, activity to flourish.

However, it has also contributed to the continued demise of other city centre streets; for example, Carl mentioned London Road, once Liverpool's chief shopping area (Couch, 2003), as ripe and ready for development:

> It's just London Road now, that's the dark side now isn't it, that needs upgrading, you've got the hospital,[27] so that will come up ... you've got the bus stop there, so that's not too bad, you've got good parking behind TJ's, you've got the tropical medicine place and the university,

ultimately suggesting that, building on an existing successful South Indian restaurant, it would make a good "curry mile". Respondents also expressed concerns about empty, iconic buildings which have an uncertain future, including the Irish Centre (Susie), the famous Lewis's building (Carl), and the Lyceum, a former post office on Bold Street (Mary).

Access to and ownership of contemporary regeneration – inclusion, exclusion, and 'ambassadorisation'

As detailed in the introduction to this book, Liverpool remains a city with problems, touching on all aspects of social life, from local authority spending cuts foisted on the city to austerity measures, to crime, mental ill health, and low life expectancy. These social ills have left the city with something of an "image problem" (Coleman, 2004, p11), a problem that respondents showed a clear awareness of. Mirroring Maggie O'Kane's post-Bulger proclamation of "Heysel, Hillsborough and now this" (*The Guardian*, cited in Scraton, 2009, p251), respondents spoke about the impact these tragedies had on them and, subsequently, the city's image.

Teresa, for example, recalled that both her husband and her father had travelled to Heysel Stadium to watch the match live, and that the trauma and shame of the event had a lasting impact:

Teresa: My dad said, he used to be a season ticket holder, and he said he'd never go to a match again after that ... they got marched out at gunpoint and he said he'd seen a lot of both [teams'] supporters, he was disgusted in the Liverpool supporters and he was disgusted in the, they were Italian weren't they?

Clare: Yeah, they were Juventus.

Teresa: That's it, and he said it was just awful, and he'd never really go into detail or anything, but he just went I'm not going, I'll never go to another match, and he never did.

In recalling the Hillsborough disaster, however, respondents displayed no sense of shame, reflecting the long-standing belief in the city, now validated as both 'fact' and 'truth' since the verdicts of the new inquests in 2016, that supporters played no role in the disaster. Sisters Pat, Maureen, and Colette, who lost Maureen's son at the disaster, displayed a sense that Liverpool fans were constructed as the cause of the disaster simply because they were *Liverpool* fans:

Clare: Do you think it would have been the same, played out the same with the lies and all of that, if it was a different football team involved?
Pat: It would depend on what football team, but …
Colette: I do, yeah.
Pat: As you [Colette] said, with Liverpool they make it worse don't they?
Clare: So if it was a London team, say …
Pat: It would have been different, yeah.
Maureen: Or if it had been rugby, cricket, or racing, race meeting or something like that, could easily have happened.
Colette: It would be a whole different attitude I would think.

Jill was the only respondent to link Heysel, Hillsborough, and the murder of James Bulger together, yet she did so in a manner very different to that of Maggie O'Kane:

Jill: We had the football side of things back in the 80s, we had Heysel didn't we, and then Hillsborough, but then that's been turned around completely thankfully, obviously sadly we had the Jamie Bulger case, that's very sad, but then again that could happen anywhere.
Clare: Yes it could, although lots of people at the time said it could only have happened in Liverpool.
Jill: Yes, but I would disagree entirely with that.

Thus, respondents were aware of established narratives, particularly outside of the city, that Liverpool is a 'criminal' city; a place where the most singular of crimes occur *because* of a set of causal characteristics which facilitate such activity – a unique breeding ground where forms of crime perceived as distinct to the city can flourish. However, respondents recognised such narratives as falsehoods, and disaggregated the Heysel disaster, the Hillsborough disaster, and the murder of James Bulger into separate, unrelated events – Heysel, notwithstanding the mixing of fan groups and the state of disrepair of the stadium, was something the people of Liverpool need to take responsibility for; yet Hillsborough is a state crime that the state has tried to deny by deflecting blame on to victims, and the murder of James Bulger, while both shocking and tragic, could have happened anywhere; Liverpool is merely its setting rather than its cause.

Relatedly, several respondents acknowledged the 'whingeing Scouser' tag, perhaps attributed to the city's populace as a result of refusing responsibility and blame for events like Hillsborough, and discussed how they felt about it. Susie, Martha, and Mary all indicated that the term 'whingeing' could, and should, be substituted with something less derogatory:

> I think to some extent it's true that we do whinge a lot, well I do, but it's not, we want what's right, and I'm proud of being a whingeing Scouser if that's what it means, because I'll never give up fighting for other people who haven't got a voice. (Susie)

> That's a load of my arse that, all the way, we will talk about what trials and tribulations we've had to put up with, but we don't cry and moan about them, we just discuss them and move forward with them, and I think we're a very, just get on with it type of people, we won't let people walk all over us, and this is probably where the whingeing thing comes from, we'll go, hang on a minute, no, and that's not whingeing, that's standing up for yourself, that's having confidence in your knowledge as being truthful, the truth will out type of thing, you know, that's not whingeing … that's resilience. (Mary)
>
> Well I think we've got cause to whinge to tell you the truth, I think nobody ever looks at why they just say oh a gang of Scousers who have whinged, well what have we whinged about, we've whinged about Hillsborough when we were right about Hillsborough … we've had a lot to moan about, they don't call people from Manchester moaners because they've not really had anything to moan about, alright they have poverty, but they haven't had injustice, that's what it is, the injustice we've had to put up with, and not just Hillsborough, but people who worked on the docks, and they closed the docks, and mass unemployment, and the way Black members of our city have been treated in the past, and even now I think yeah, we have had a lot to moan about … they don't say that we've got strength, or that we're the type of people that battle for justice. (Martha)

For many of the respondents, the 'whingeing Scouser' label is enmeshed with other stereotypes about, and judgements on, Liverpool and Liverpudlians. This is encapsulated in Joey's account of an experience which made him feel "really angry":

> I heard someone talking about Liverpool and let's just say bad mouthing the city and Scousers. I just stood with this group of people listening to this nonsense these two particular people were spouting, and what's worse, most of the group were agreeing with them. I let them go on and waited for my time then just said "finished now? Has anybody here actually been to and experienced Liverpool?" The group just fell quiet and not one person said they had been there, [I said] "oh there's a surprise, you should really try to make educated comments, do try to keep up, because right now you look really foolish, and where are you from, oh let me guess, Shitsville, that must be where people who speak copious amounts of shit are from".[28]

Themes of fairness, justice, and truth are relevant to the city's reputation then, and can be understood as the basis of a circular and self-fulfilling pattern; when something negative occurs towards the city and/or its children, the city (a) acknowledges the harm that it has caused and (b) seeks redress. This then leads to the application of the 'whingeing Scouser', 'self-pity city' label, which is subsequently disputed and challenged by Liverpudlians, further entrenching the view that the city is self-absorbed, defensive, and angry. Such processes and patterns are deeply frustrating to those for whom Liverpool is home, who simply want to be viewed with parity and not be pigeon-holed or condemned, as illustrated by Mary:

> We've had, what, thirty, forty years of being perceived as a gang of horrors, basically, and I suppose like the old saying, if you tell a dog to meow enough it will start to think it's a cat, but we've never thought we were a cat, we've carried on barking!

Perhaps one of the reasons that Liverpudlians feel so affronted and embattled by negative stereotypes and labels is the widely held belief that Liverpool is a friendly city with a warm, hospitable, and affable populace. Several of the respondents focused on the friendly image of the city, for example Sandra:

> [Liverpool people] are friendly with a great sense of humour, we are by far the friendliest people on earth. We are very lighthearted and have a good sense of humour, we take care of our own and are always there to help, genuinely helpful and honest people.

For some respondents the inherent friendliness of the city manifests itself as a welcoming spirit, ready to embrace all, leading to an 'inclusive' atmosphere, as indicated by Mary:

> I think especially because of where I live in Liverpool, and the work and the community, I can see that the Scouse sort of welcoming and openness, it draws people of a like mind from other places, because I see in school,[29] you know, people from such a wide range of backgrounds all over the world, literally all over the world, and the majority of them are open to this sort of community feel, this warmness, this openness, this kindness that you do find, and sort of take it on board and carry it on, and I think if they came somewhere, to another city that wasn't quite as open and welcoming, then *they* wouldn't be open and welcoming.

However, respondents consistently expressed concerns about processes of inclusion and exclusion within the city, and several ways in which inclusion, in the backward-facing, heritagisation climate, is not something guaranteed for all. Joey, Paul, Peter, and Sandra all spoke about the importance of free entry to, for example, cultural buildings and events, and the danger that 'ordinary' Liverpool people may be prohibited from engagement due to high costs. Several spoke about access to leisure, particularly night life, and how access has changed over time and in response to changes in the city's cultural reputation and standing; Jodie, for example, commented on the increased commercialisation of the Liverpool night-time scene, cultivated to attract tourists:

> I don't do Mathew Street, I just don't like it, I don't like it when they're all commercialised, I feel like I'm in Tenerife because everyone's touting you

> to go in and have a drink, everyone's completely rotten drunk, I don't like that kind of way, I think the ale's cheaper, it just tastes nastier. It just reminds me of a cheap night in Benidorm really … it's all hen dos and stag dos and whatnot, I don't see how you enjoy going to a different city for a night out because you don't know where the fuck you are.

Mary, Susie, and Teresa also expressed concern about hen and stag parties in Liverpool for the weekend, particularly in relation to conduct; however, Ronnie noted some positives regarding the phenomenon:

> I don't know a thriving city where that doesn't happen, you go to Dublin, to Barcelona, there's people in terrible clothes with pink sashes on by the end of the afternoon, that's kind of what happens, I think the places where it doesn't happen probably would like it to. Stoke would kill you for a hen party.

Peter suggested that, in spite of the proliferation of out-of-town revellers, there is still room for everyone and "working men and women can still go to some places in town"; but equally expressed concern about *who* leisure and night-time consumption are aimed at:

> Well to be honest with you I mean it's nice for things going on and stuff like that, but I don't really think it benefits the people of Liverpool, when I say the people I mean the people, the masses of the ordinary people who live everyday all over the city in different areas, I don't think they'd benefit from nothing like Liverpool One, and the Albert Dock, so it's lovely the way they've renovated it all, and there's lovely shops and cafes, but the price of them down there, there's not many people who are in low paid jobs in Liverpool, or who are unemployed in Liverpool who can go down and socialise there, there's lots of hotels all along the waterfront and on the outskirts of town, they're popping up everywhere, and it's, basically it's for other people, it's for people with money basically, I'm not knocking it completely, but I'm just saying it's not beneficial in any way to the people of Liverpool in that sense.

Susie spoke about a social leisure event which is viewed by many as essentially Liverpudlian: The Grand National.[30] Like me, Susie and her family live very close to Aintree Racecourse, and she spoke at length of her feelings about it, particularly during the Grand National race meeting every April. Her testimony encapsulates the ways in which a local Liverpool 'culture' has been co-opted, commercialised, and ultimately used against those who regularly attend, particularly the women:

Susie: I didn't think the Grand National would have as big an impact on us as it does, when we first moved in it was alright, it wasn't as commercialised as it is now, you'd get a couple of tickets, never went because I don't believe in it, but I'd think ok you know, and it didn't cause us that much problems, and now it is horrendous. We've got the no parking in our street from Thursday to Saturday.

Clare: So where do you park the car?

Susie: We have to put it in my sister's in Netherton, and she'll run us back, then they come down the street thinking it's a quick way out then realise it's a dead end, people pissing all over the place, pissing on the step, dreadful, just hate it … . I don't like all the people being sick and drunk, and all that kind of thing, I don't like, obviously for the violence against the animals, hate that.

Clare: That's why I won't go.

Susie: And I think the other thing now, it's a bit of a commodification now, cultural appropriation, because it used to be a working-class thing going to the races, and all the Scouse girls on the Thursday[31] used to get dressed up for Ladies Day, and all those things, and now we're priced out, same with the football.

Clare: And if you are going, you have to start basically saving up from Christmas don't you?

Susie: Because of all this vilification in the press, oh it's awful, slut shaming, fat shaming, and all kinds of other shaming, it's vile, it's absolutely vile, and I think it's awful, and that bloody Queens pub, have you gone in and seen all the pictures on the walls in there?

Clare: No, I haven't been since it's been done up.

Susie: You need to go in just to see the pictures on the walls, so what they've got is the worst of the worst pictures all framed on the walls all in the pub … you can't move, literally your car's taken away, and there's no buses, and I've got limited mobility, I'm waiting for a new knee, it's horrible, just horrible.

Susie's account of Aintree life during this race meeting touches on several issues pertinent to both the Liverpool home and the Liverpool reputation. Parking restrictions and police barriers[32] can inconvenience residents, and alcohol consumption at the event is a causal factor in antisocial behaviour, including public urination in streets and on people's property and, less frequently, vomiting, defecation, and sexual activity, creating a jarring dichotomy between the elegance and opulence of the event, and the unsavoury conduct of some racegoers. In this sense, Aintree races can have a detrimental effect on residents' immediate experiences of home.

However, as indicated by Susie, it is constructions of, and responses to, racegoers that have a bigger, more upsetting impact. In recent years, the media have begun to highlight the dress, demeanour, and conduct of racegoers, particularly women, and processes of 'shaming' serve to castigate such women for attempting to access what has historically been the domain of the middle and upper classes – the more select aspects of race going –-champagne, fine dining, elegant clothes, and elaborate headpieces. The media appear to mock Liverpool women at the races on the grounds of both their gender *and* their social class, as if they are trying to access a way of life that their assumed class position will never allow them to. This is reminiscent of Skeggs' (2004) conceptualisation of working-class women socialising; how they are read by the wider world, and how their access to, and enjoyment of, social arenas is limited by popular negative notions of class, taste, and femininity – summed up in the attachment of the given names 'Sharon and Tracey'. Thus, "class divisions interrupt both participation in one's own consumption, the consumption of others and in being consumed" (p172). Constructions of Liverpool women simultaneously impede their own enjoyment of the races, and add to the perverse enjoyment of others as they look down their noses at working-class women trying and failing, as they see it, to be 'classy'.

Susie's testimony regarding the Grand National meeting is also pertinent to the dichotomy between inclusion and exclusion because of her health. At the time of the interview, Susie was waiting for knee replacement surgery, and therefore had limited mobility and found walking even relatively short distances difficult. However, she had no choice during the races:

Susie: When I come home from work, if I get off the train at Aintree [Station], you've got to go all the way round the world haven't you?[33]

Clare: So the coppers never let you through the barrier?

Susie: No, so you've got all that extra walk haven't you, and you're fighting the rising tide of people that are coming out who are all drunk and everything else, so you're swimming against the tide, and I can't walk that far, it's just too far.

Susie was not the only respondent to cite health and/or disability issues as an impediment to inclusion in Liverpool. Rita, who suffers from chronic obstructive pulmonary disease (COPD), often needs additional oxygen and, subsequently, leaves the house infrequently; the only time she has been to the city centre in the last ten years was to a family wedding at St Nicholas' Church, close to the waterfront, and she has never visited Liverpool One. Jill also has limited mobility due a chronic condition, joint hypermobility syndrome, with which she has suffered since she was a child, and limitations on her mobility cause her to see developments like Liverpool One as exclusionary:

Jill: Generally speaking now, not only just from my own, that particular perspective, if you look at Liverpool One, or the city centre, how many places are there to actually sit down? If you're elderly, or disabled, count the number of benches or chairs to sit down, unless you're going to a café or a restaurant.

Clare: Yeah, lots of people sit on those sandstone steps don't they?

Jill: Yeah, if you're disabled or elderly, even I, getting to sit down on a step now for me is hard, so from a more practical [point of view] I don't think it's as inclusive as it could be.

Gary spoke at length about the exclusionary nature of the city centre that he has encountered since losing his right leg a number of years ago; an issue that he has raised verbally and in writing many times with his local councillor, his MP and the current Liverpool Mayor Joe Anderson. Gary's disability permeates virtually every aspect of his relationship with public space and social engagement in the city, which in turn impacts on his sense of self, his self-confidence, and his masculine identity. For Gary, neither Liverpool City Council nor businesses who operate in Liverpool take adequate account of the needs of the physically disabled, whether in terms of access, parity of treatment, or his physical safety as he makes use of the city centre:

> As an amputee, first of all the parking, parking is horrendous, but the biggest one for me is shop access, I know the equality access they had to make reasonable adjustments, is it reasonable for them who have two legs and don't use a wheelchair, or is it reasonable for me, who has one leg and uses a wheelchair? There's that many broken pavements you've actually got to watch because if you hit a broken pavement you're out of your wheelchair, and there is an opportunity of you falling into the road … you've got the lip [of the pavement], so you've got to try and flick your wheelchair up, keep the momentum to go up, now because I've got good upper body strength I can do that, but not everyone is as blessed as me. I should just be able to roll wherever I want, I fully understand how it's an old town, old city sorry, built on hills and whatever, but it could be made better.

Gary has encountered many examples of bad practice, ranging from being told he would have to be carried into the HMV store, "the security guard on there at the time said, do you want me to lift you up? I said, do you want me to knock you out?" to enduring the elements at Anfield football stadium:

> When *you* buy a season ticket, or a ticket for a football game, you buy a seat, *I* bring my own, so all I'm paying for is a seat they can't do anything with, they can't put seats in because they wouldn't be able to see over the wall, so they're not giving me anything, but I'm paying the same price as everyone else … it was my choice to go in the Anfield Road end because I prefer it, but in the winter if it rains the rain drips off the roof, and Liverpool Football Club give you a poncho in a bag that looks more like a condom, it really does, or to put over your knee they give you a black bin bag, so you're paying £34 but this is what happens.

However, he has also encountered good practice, sometimes in the least expected places:

> The Cavern[34] [has] stairs going down; now we had friends come up from Portsmouth that we'd met on holiday, well they wanted to see Liverpool, so as we're going out, it's Saturday afternoon, and I said "there's The Cavern, do you want to go in?" He said, "oh what about you", I said "no", then the two bouncers said "do you want to go in mate", I said "well I'm with these", he said "come on we'll sort you out" and he took me round Cavern Walks round this dodgy little passageway in the back, in a lift, took me down and got me into The Cavern … that's going above and beyond, you don't mind

> that one bit, but when you've got a newly built shop and you're thinking how do I get in here.

What upsets Gary most though is the deliberate construction of old-fashioned, antiquated spaces, consciously designed to complement existing built structures that are deemed to have historic value. This heritagisation of public space is most apparent to Gary at the waterfront:

> Where the new Beatles statue is towards the terminal it's nice and flat, but the road is then on a camber one way, and on a camber the other as you come back round by the Radisson, but do-able in a wheelchair, it's a challenge, but do-able. When you come back the other way, around the front they've got one tiny line of pavement, but you've got these big wonderful cobblestones, in the 21st century do we need cobblestones? It's wonderful, the heritage is wonderful, I'm all for heritage, but surely practicality has got to take, you know, why not put the cobblestones way over to the left, so you can see what *used* to be cobblestones, isn't it dangerous for people to walk on cobblestones?

Gary's frustration with his access to public space, and his safety when he does so, reached a peak on Armistice Day a couple of years ago; as an ex Royal Navy officer, Gary attended the service of remembrance held at St George's Plateau; however, the built nature of the venue led him into trouble:

> So we're there on St George's Plateau, and it was [wife's] uncle that was going to push me, I didn't need pushing, but it looked better for the parade, so they called the parade, sat up in the wheelchair like that as you do, he went by the left, quick march, [the wheel of the chair] got stuck in the cobbles and he just tipped me out, so I'm like that [lying prostrate] and a few other lads marching past, I can see the humour in most things Clare, there was nothing broken, the only thing that probably hurt was my ego, my pride, but I couldn't stop laughing, some of the general public saw it as well.

St George's Plateau appears to be Liverpool City Council's designated and approved space for public gatherings, as indicated by its intention to create "a new event space" at the site under the Better Roads City Centre Connectivity Scheme (Liverpool City Council, 2016), almost as if public gatherings are only validated as legitimate and appropriate if they are held at this site. Gary's testimony indicates again that issues of inclusion have not been given enough thought, and that heritage takes precedence over real inclusion and public safety.

An identifiable group that respondents felt were subjected to social exclusion was homeless people. Several respondents noted that street homelessness in Liverpool city centre appeared to be worsening rather than improving, in that more and more people are visibly living in public space, for example, Susie and her husband both have experience working and volunteering with homeless people, and note that

> You'll see it as you go out now, a massive increase in homelessness, *massive*, and I can tell you this, this time last year [October 2015] you'd see on average nine to ten people a night on homeless outreach, 47 they saw on Sunday night, and that was in an hour and a half, *an hour and a half* Clare.

Gary, although very well versed in inclusion issues in terms of disability and gender,[35] displayed some stereotypical views about street homeless people, as indicated in this exchange:

Gary: Is this what you're putting in your doctorate?

Clare: Yeah I will, so one of the things I've been focussing on is access to space in the city centre, and access is denied in a range of different ways, so obviously something like disability, but not just that, so like, for example, homeless people, if someone sits down outside John Lewis in Liverpool One they just get moved on immediately, like that, off you go, because it's all private land.

Gary: Ok, what do you think about homeless people?

Clare: I think there's a lot more than there were before.

Gary: Where are they from?

Clare: Well, do you know what, if they're not from Liverpool they can't access the services, did you know that?

Gary: Well they can't access the services, because they're all Eastern European, that's the point I'm making, we allow them, or they're allowed into Europe up until two or three years ago.

Clare: Well it depends, Romania and Bulgaria were two years ago I think.

Gary: Yeah, so they're the ones who are mostly on the street, but they're getting off it and going to houses, and you see them begging and they go round the corner, take the old coat off, and get into a BMW.

However, all the other respondents who discussed street homelessness did so with a sense of compassion and concern, for example, Tony, who questions whether money is spent wisely in the city:

> They [Liverpool City Council] can spend it [money] on a million better things than putting a load of giant puppets on for people,[36] when I work in town at night time, you see it of a day, but you don't see the full scale of it till the night time with the amount of homeless people dotted all around our town centre, it's terrible, and we're supposed to be kind natured people, which we are, but a lot of people don't see it because you go home and you're with your family and stuff, but it's unbelievable, and we always speak to them, get them a cup of tea and that and some of them, well they're all normal people, but they've had different problems, I don't get why we've got big buildings

> all boarded up with nobody in,[37] help them people [to get in] somewhere, it's not hard.

Here Tony has hit on a clear irony in the way the city is perceived and marketed, and the way it conducts itself and makes decisions about spending priorities – Liverpool is constructed as a 'friendly', hospitable city, whose children have hearts of gold, yet the way many in the city respond to homeless people, from individual residents to elected council members, contradicts entirely the friendly, welcoming image that the local authority and its collaborators in the private sector project to the wider world. To put it another way, the city opens its arms to tourists and travellers who will spend money in the city and promote its best side to the rest of the world, but is decidedly closed up when it comes to those in need. This is apparent, for example, in Susie's testimony from her husband's voluntary work with rough sleepers in Liverpool, and the anecdotal evidence that Liverpool City Council does not hesitate to invoke the Housing (Homeless Persons) Act 1977, which affords local authorities the right to move rough sleepers on unless they can prove a "local connection"[38] (Dwyer et al., 2014).

This however is not to say that the people of Liverpool are incapable of compassion towards excluded groups like rough sleepers; as discussed, Susie's husband volunteers with homeless people every Sunday night, handing out hot food, warm clothing, and so on; a school friend of myself, Mary, and Jackie does the same on Mondays; and Mary's nephew does the same on Tuesdays. Despite the popularity of 'urban myth' type ideas held by Gary, there is a great amount of compassion, political support, and practical help given to homeless people and other excluded groups at the agentic level, acting as both a counterbalance and a challenge to the structural-level drive to eliminate rough sleeping and problematise rough sleepers. For Susie, excluded groups like rough sleepers are denied status as citizens of the city with rights and freedoms:

> They're non-citizens aren't they, they're seen as not being like part and parcel, so you've got the elite coming in and buying all these riverside properties, and all these new apartments and stuff, and probably never living in them, and then you've got people who are on the streets in dire need who've got nowhere to go … . I think it's great that we have people coming into the city to live and work, absolutely, but not at the expense of the vulnerable … everyone should have the same citizenship and those rights, and they're not, they're being knobbed off basically, and not having access to all the things they should have, well no, I don't think so.

In this sense, the inhabitants of Liverpool can be understood as divided; categorised either as residents of whom the city can be proud, and who should be showcased to the wider world, or residents who should be less visible, whether due to a disability, or due to more complex social circumstances leading to rough sleeping, street drinking, and any other forms of public 'incivility'. To put it another way, reminiscent of Hollander's (1991) contention that the term 'home' is more respectable that the term 'house', which had come to be used as a short version of 'whorehouse' – respectable Liverpool people are enfranchised to call the city their home, while those deemed unrespectable are not afforded the status of home, as they are seen as undeserving. Against this contextual backdrop, Gary's views on street homeless people, or at least their root cause, become more understandable – primary definers seek to exclude various groups on various bases, and classic divide and conquer tactics appear to have turned Gary, as a disabled man, against homeless people who also have their access to the city restricted.

The issue of pride in the city, pride in the city's inhabitants, and the Liverpolitan duty to present Liverpool's best image to the wider world, was one that came up several times throughout the interviews. In the majority of interviews, and all of the questionnaires, I raised the idea of Liverpool people as being ambassadors for the city, responsibilized into promoting the best aspects of the city, its populace, and its diaspora – this idea came directly from an event I attended in June 2015,[39] at which Joe Anderson, Liverpool's elected mayor, gave a short welcoming speech. During the speech, Anderson spoke of Liverpudlians as 'ambassadors' for Liverpool, representing the city to the wider world and doing the city proud. Respondents who commented on Anderson himself universally did so negatively, with criticisms ranging from his appearance to his approach to local politics. Responses to the idea of ambassadorisation, however, were more mixed. Enid, Joey, Paul, and Sandra, who all left the city to live elsewhere some time ago, were very much on board with the notion:

> I definitely see myself as an ambassador for the city. (Enid)
>
> I think Liverpool people have a responsibility to emphasise what's good about the city ... if Liverpool is to continue to develop it needs people to talk more positively about the place. (Paul)
>
> I definitely do see myself as an ambassador for Liverpool and always try to adhere to the reputation we have as being genuinely helpful, honest with a great sense of humour. (Sandra)

Jill offered an explanation as to why ex patriot Liverpudlians express such strong positive views towards their home city:

> Obviously they've got this very, it's an imaginative construct basically, it's the Liverpool of their imaginations, it's not the real Liverpool, because if they've not been here for so many years then it can't be the real Liverpool, it's very much based on nostalgia, and on imagination.

Dave, who has lived away from Liverpool for over 20 years, and currently lives in Ireland, does not see himself as an ambassador for the city, stating that "my relationship to Liverpool is that of an ex patriot. It's where I come from but not where I belong". This raises questions as to why the majority of the respondents who are part of the "Liverpolitan diaspora" (Heseltine and Leahy, 2011, p43) embrace Anderson's message of ambassadorisation, yet one, Dave, does not. The answer may be that Dave displayed the least positive view of Liverpool in responding to the questionnaire; whilst the others speak with great affection about the city and its children, Dave is more evenhanded – acknowledging the city as his birth place and no more, and suggesting that the city is a place like any other. There is none of the pride and sentiment that comes through in, for example, Joey's words:

> Liverpool will always be home to me, I love the city, in fact when I'm talking about the city I call it 'my Liverpool' ... to me Liverpool is the best city in the world ... it's a feeling that you have when you're from Liverpool that you cannot put into words, Scousers will understand this.

Dave is dispassionate, taking a more analytical, detached view,[40] for example:

> I think it's great that the city has been cleaned up and is now attractive to tourists. For me there is a sense of loss of the 'old world', the working-class history in the Victorian buildings, terraces etc., they have disappeared along with any evidence of the existence of the city's true inhabitants. I also feel that this is a façade which gives the illusion of wealth but there seems to be no infrastructure for the residents to benefit from any of it. I would like to know where industry fits into this regeneration.

It would be wrong to speculate about Dave's stance on his home city, and possible reasons why, beyond what he has willingly provided. What can be concluded though is that Heseltine and Leahy's (2011, p51) suggestion that

> Many of these people retain not only relatives or friends in the city region but possess an abiding loyalty and affection for the place. Perhaps more so than for some other places, Liverpool seems to be a place that its sons and daughters retain a passion for.[41]

is challenged and undermined by testimony like Dave's.

The sense amongst respondents still living in the city tended to be that all Liverpudlians, by being good citizens leading lives to be proud of, act as 'natural' ambassadors for the city without being called upon to do so:

> I'd like to think that when I go anywhere, or most people I know go anywhere, I'd like to think that we are sort of ambassadors for our city ... I'd like to think myself I'd be a good ambassador, but for him to be saying we should be ambassadors is a bit condescending I think. (Peter)
>
> You should definitely represent your city properly if you're out and about or whatever, yeah definitely, you shouldn't have to be asked. (Tony)

Jill, however, was positively infuriated at both the suggestion and Anderson himself:

Clare: One of the things he [Anderson] does say, I heard him say this with his own gob about 18 months ago, he said that Scousers are, and should be, ambassadors for the city of Liverpool. What do you think about that?

Jill: Well I mean does everyone everywhere have to be an ambassador for their city, so if you come from Manchester, I mean it's quite frankly a load of bollocks, he's in the job, he's getting paid a nice, fat salary, he should be the ambassador, that's his job to be the ambassador, and people like him in the council, not me, and not Joe Bloggs from down the street, alright pay me a nice, fat wage and I'll be a bloody good ambassador … they're getting paid x amount of money for being an ambassador for Liverpool, therefore in a way that's their job, I'm not getting paid any money, and neither will everyone else in Liverpool, you saying then about ambassador, alright give us all a bloody good wage, everyone in Liverpool a damn good wage, the wage they're getting, then we'll be ambassadors, expect us to do it for free, I don't think so!

In keeping with her views on excluded groups, Susie felt that Joe Anderson's call for ambassadorisation only applies to *some* Liverpool people and, therefore, is inherently wrong:

> The city means different things to different people, and if we can't all be ambassadors why should some? So like for example he wouldn't be asking the Polish diaspora who live here to be ambassadors would he, he would want to exclude them, the street drinkers, the sex workers, all those populations, they wouldn't be ambassadors would they, and they wouldn't be citizens, that's what gets me.

In this sense, and taking on board Kidd and Evans' (2011, p764) claim that "home is the place where you feel accepted", having Liverpool as one's 'home' is a status that one has to acquire, *earn* even, before feelings of acceptance can develop, and before the mantle of 'ambassador' can be attributed. The status of home, then, can be denied to those deemed not to deserve it. For those not worthy of 'home' in the Liverpool city context, *fremde* – the sensation of being an unwelcome guest in a hostile place – is arguably more applicable than the sensation of nationhood and belonging referred to as *heim*.

Jill, aside from her view that ambassadorisation is "an absolute cheek", applies a 'responsibilization' thesis to the concept, and subsequently finds it even more objectionable:

> I think this idea of Liverpool being special in a way is a disadvantage to people from Liverpool, because if the onus is on you then to be a bloody ambassador for Liverpool you know you've got this responsibility, and how are you going to be able to just get on with your life in a normal, and natural way … that responsibility is I think actually for the people of Liverpool is a

> negative, they're putting all that responsibility on you as an ordinary person of Liverpool who's not getting paid any money for actually doing the job, to do the job.

Here Jill equates the call for ambassadorisation of the 'good' people of Liverpool with Foucault's (2008) governmentality thesis and Garland's (2001) subsequent development of the concept of responsibilization – within a neoliberal, 'statecraft' context, individuals are burdened with the responsibility of furthering the capitalist project, through appropriate conduct, activity, and beliefs. In other words, individuals are expected to self-govern; those unable to do so are problematised and excluded. Just as people have been reconstructed within the neoliberal project as having full responsibility for themselves and their families, and therefore to blame for any misfortune, the reputation and status of the city now also rests on the heads of individuals rather than any socio-economic process. Respondents' acceptance of ambassadorisation, albeit well intentioned and honourable, can be understood as the result of hegemonic processes, gently and subtly coercing compliance from good people who simply want the best for their home city. Equally the concept of habitus, as explored by Bourdieu (1985), is applicable here; regenerated Liverpool has developed a 'field' of 'rules' that the 'children' of Liverpool must successfully negotiate in order to evolve from Liverpudlians into Liverpolitans – entitled to a stake in the city, entitled to view contemporary Liverpool as their home, and worthy of the status of ambassador. What can be concluded, then, is that in terms of respondents, those who live away from the city (with the exception of Dave) are more amenable to taking on the mantle of ambassador; those who remain in the city tend to see the suggestion as patronising, divisive, and exploitative.

Bread and circuses: Event-driven regeneration

Dave's query around the role of industry in contemporary Liverpool regeneration hints at a further key theme that has emerged during the course of the project – the focus on events; ephemeral and transient – as opposed to lasting, permanent developments to infrastructure. 2008, ECOC year, was essentially an 'umbrella' event holding together a series of events, and marked the entrenchment in Liverpool of the belief that single events, or a series of events, can bring benefits that extend beyond the end of the event; benefits that can continue to have a positive impact on the city and its residents long after the event has finished. The lesson that Liverpool City Council and its partners appear to have taken from 2008 is that commemoration is the way forward, as a series of events have been concocted to mark the passage of time since a particular occasion; some of which, for example 2017's '50 Summers of Love', a "season of events inspired by the golden anniversary of the summer of love" (Culture Liverpool, 2017), seem spurious, as if the city is grasping at anything that might plausibly be memorialised. Starting with ECOC, this section explores participants' memories and feelings about such events in Liverpool, and the impact they have had, and continue to have, on the city and its inhabitants.

All of the respondents had an awareness of ECOC year and, in the main, all viewed ECOC as an accolade for the city, and something to be proud of:

> I was really proud because I thought it was going to bring like tourism and people, and they can see the real Liverpool, and the nice parts, they [the media] tend to pick these awful ones out, so when the Capital of Culture, I thought it was just great for the city to be honest. (Teresa)
>
> I think it was a very good thing, yeah I think the investment that went into the city because of it, and I think it really did put it on the map. (Martha)
>
> In 2008 when Liverpool won the European Capital of Culture it really put it on the European map – it also put us on the world map for other than the music it was known for. It opened a lot of eyes to the fact that there was so much more going on in Liverpool. (Enid)

However, several respondents remembered very little of the ECOC year, and struggled to recall specific events, like Christine; "Do you know Clare, I can't, I can't remember anything about it really", and Rita; "I can't think, I can't, how long ago was that, I can't remember". This could simply be a reflection of the length of time that has passed, or a result of the inherent temporary nature of events – they start, and then they end. Difficulty in recalling ECOC year and subsequent events might stem, though, from the limited extent of engagement reported by respondents.

Christine remembers seeing *La Princesse*, the mechanical spider and first of the 'giants' to be paraded around the city centre, but she happened upon it accidentally:

> I remember going into town and I was with [husband], it must have been an afternoon, I don't know whether we got the train into town, must have got the train into town, and you know that pub at the side is it the Crown? It was on the side there, and I remember looking and going, my God the size of that!

The other events that she could recall, the other 'giants' and the Three Queens, she watched on television, as "I'm not particularly fond of getting shoved and pushed in crowds now". Anne and John did not engage, largely due to John's temperament:

Clare: Remember when it was Capital of Culture year?
Anne: Oh yeah.
Clare: Did you do anything to join in with that?
Anne: No.
John: No.
Clare: How come you never went? Just weren't interested in it?
Anne: Oh God no, I was interested, I watched it on the telly and that, it's just getting there to be honest.
Clare: Palaver?
Anne: He [John] wouldn't go to that, or like that would you, I can't drive, so I get the lift off him.

Clare: Is it the crowds and all that, is that what puts you off?
John: I don't like crowds me, I can't even wait at a bus stop.
Anne: [He] can't even wait in the queue when we're in the supermarket, if someone is fumbling with their purse at the checkout … very impatient.

Rita recalled the 'giants' when she was reminded of them:

> Oh yeah! Well we didn't go on Capital of Culture, we saw the giants two or three years ago wasn't it, she went along, they went along Kenny [Kensington high street], and they took me up to see that, it was good.

Significantly, Rita only agreed to be taken to see the 'giants' on their last visit in the summer of 2014, when the decision was taken to extend their route to take in some of the inner areas, including Kensington; she was comfortable engaging with them a short distance from her home, but not in the city centre.

Martha and Jeni recalled going to see the 'giants' in the city centre, both finding it enjoyable and moving:

> The spider, especially the spider, I remember that happening and taking my niece, and I just thought the amount of people that were out, and the money that would have been made, and everybody in the country was talking about us. (Martha)
>
> The giants with the uncle, I loved that, I found it like dead emotional just looking for her uncle and they were separated and everything, I must have been pregnant or something at the time … I just thought it was lovely like the story behind it and everybody going to see them, cheering them and everything, it was just all dead nice. (Jeni)

Further, several of the respondents commented on the importance of events like the 'giants' being free, recognising that cost can be a barrier to engagement as "there may be elements of the community that can't afford to do certain things" (Joey).

Equally though, there were respondents who were nonplussed by such events, displaying cynicism and lack of interest. Jeni's husband, Tony, did not join her when she went to see the 'giants', stating:

> It was alright, it looked like a good event to go and watch, if you want the truth it was just alright, not for me though, a good little spectacle, but not for me, not my cup of tea, it looked alright though, I'd have had a look if it weren't so busy I think.

Peter had similar feelings, and queried the ability of events like the 'giants' to enhance everyday life in Liverpool:

> It's a good day out and stuff like that, but it shows the people of Liverpool, on the whole they're really good people, they're great and they love having

a good time, but they're easily pleased at the same time, and if they see this big spider called whatever her name is, and these giants are a lovely sentimental story on a summer's day, they're always going to look at it because it's something to do.

Echoing Tony's comments about how money is spent within the city, Peter feels that money spent on, for example, the 'giants' could be put to better use:

Yeah they have a great time for that one day and all that, but the money it costs, when we could be putting that into other stuff like computers for kids in schools, or stuff for in hospitals, or Alder Hey [children's hospital] for things, money to create jobs and houses, or anything for the homeless, it's great for the day out for people to enjoy all that spectacle, but in terms of paying for it through the council and people's money coming out of it, I think there's far better things that should be getting paid for, yeah, most definitely.

Further, respondents expressed concern that money spent in the city during such big events may not stay in the city, as summed up by Roy:

There's a girl from Leamington Spa I know, she said I'm coming up this weekend, booked the hotel, and she came up for the whole of the princess (spider 'giant') and saw it all, and she said do you know what, the weekend cost me about £400 just in accommodation and things alone, she said but I've never had such a great time, she said I ended up down Mathew Street doing this, doing that in Liverpool all over the place, so I'm thinking she must have spent about £600, so it's had a financial benefit to the city, I'm just hoping it didn't all go to massive chains like Subway and Costa and all the rest of it.

Roy's concern is in line with Dave's suggestion that an "illusion of wealth", created by events like the 'giants', hides the poverty and social problems that persist in the city, and wealth accumulated during such events does not stay in the city, but is syphoned off by the corporations who have a stake in the city as a return on their investments. In this sense, events of this nature can be conceptualised as a distraction designed to deflect attention away from continuing difficulties in the city. For Roy, this is a good thing in itself:

It was very forward thinking to bring culture into everyday thinking, people went oh my God they're going to be walking around with 40 foot puppets, what's the point in that, how much is the cost, you're getting rid of buses, you're doing this, you're doing that, and yet the feeling it brought to the city, wow … instead of being a typical downtrodden Labour party[42] which seems to sort of relish the downtrodden man, how we've always got to help the lowest people up, well how about celebrating things for a change, and doing good?

Ronnie employed the Roman phrase 'bread and circuses' to illustrate a view similar to Peter's:

Ronnie: I don't agree with the bread and circuses approach, the second time the giants came around I couldn't even be bothered going to see them.
Clare: Ok, can you tell me why you don't agree with it?
Ronnie: I think it makes it appear that everything is great, my objection isn't about the money actually, everybody says oh you could have built half a hospital with that, but it wouldn't have happened … . I know people say oh well that generated x million pounds of business in the city centre that week, but it's a gloss, it's a show, it really is give the people bread and circuses thing whereas I'd rather teach people how to bake bread, I want to change every day, not wait till we have a big do one summer, so I am a bit, I suppose there's a bit of Marxism left in me.

Roy, Peter, and Ronnie are essentially discussing the same phenomenon from different political perspectives here. All acknowledge that spectacles of this nature are 'lightweight' in their contribution to regeneration in the city – they are time limited, memories ultimately fade, and the revenue they create does not necessarily go where it is needed. For Roy, the simple optimism and carnival atmosphere that, for example, the 'giants' evoke is valuable in and of itself, while for Peter and Ronnie it is a sideshow, a distraction, a smokescreen hiding, even denying, continuing negative processes impacting on Liverpool and its residents – to appropriate Mary's term, it is "fur coat and no knickers"; superficial and neglectful of basic needs. Ronnie makes overt reference to Marxism when sharing his views on events like this, assessing them to be conducive to false consciousness, whereby such spectacles are offered to ordinary people to ensure their compliance and collusion with a greater capitalist, and, in this case, regeneration goal, as outlined by Thompson[43] (2015). In this sense, spectacles like the 'giants' or the 'three queens' are little more than crumbs from the table.

The push towards commemoration, memorialisation, and festivalisation of past events facilitates a range of processes for parties engaged in neoliberal statecraft in Liverpool. Events of this nature draw big crowds; crowds who spend money, thereby increasing the return of those who have a stake in the city's revenue stream, whilst also promoting the city as a site where further events can be held, creating a potentially ever-increasing cycle of wealth accumulation. Smaller, independent retailers and restaurants may also benefit from this but, more to the point, an *image* of wealth is created, which presents an inaccurate portrait of a successful, flourishing Liverpool to the wider world – if the city can host lavish, free events, there must be cash to spare. Events like this are also constructed as tangible evidence that Liverpool is committed to regeneration of the city for its inhabitants, a cultural regeneration which may not have any significant impact on the everyday lives of the city's residents, but regeneration nonetheless; thus, Liverpool City Council and its partners are visibly 'delivering' regeneration to the people of Liverpool, in a highly ostentatious manner so that the wider world *knows* that Liverpool is regenerating.

Further, events of this kind make people happy; they provide a good day out which costs them nothing in the first instance (but may cost more than it should in terms of refreshments, souvenirs, and so on), and fosters civic pride in the city, a localised patriotism for a city which has so much to offer the world. Finally, the beauty of the past is that it never goes away; it is always there, reliable, tangible, and certain. The future however is not guaranteed; it is uncertain, unpredictable, and out of our grasp. Developing a regeneration ethos focused on the past provides a cast iron source of things to regenerate; everything that we have ever had, or ever done, is available for commemoration, an infinite resource just waiting to be mined. The future may not belong to us, but the past most certainly does, and the "civic peacockery" (Boland, 2007, p1028) of event-driven regeneration allows us to exploit it to the full.

The identifiable difference between understandings of home in Liverpool in earlier regeneration contexts – as identified in Chapter Six – and more contemporary understandings of regeneration and home in Liverpool discussed here, can be synopsised very simply: Ideas and viewpoints on *past* regeneration and home prioritize the *micro* level (dwelling, district, community); ideas and viewpoints on *contemporary* regeneration and home prioritize the *macro* level (the city centre and the city's wider, global status). This contention, and the suggestion that Bourdieu's (1985) concept of habitus can be employed to explain and understand these differences, is explored in Chapter Eight.

Notes

1 1981 also saw disturbances in Cantril Farm similar to, and in the wake of, those in Toxteth (Beckett, 2015).
2 Martha's memory lets her down here; Heseltine's first visit to the estate was in 1981 (Crick, 1997).
3 Martha continues to feel a strong sense of gratitude to Heseltine: "I was listening to Radio 4 the other week, and he even mentioned Liverpool, and what a disgrace it was when he went, and how he was proud of the investment, and I've never written to anyone in my life, and I felt as though I wanted to email him and say you've no idea what a difference that made".
4 Constructions of Heseltine as the 'saviour' of this part of Knowsley are, to a degree, ironic; Crick (1997) notes that Heseltine had responsibility for the Merseyside region in the re-structuring of local government areas in the early 1970s, and Ronnie shared an anecdote regarding Heseltine's role – "it was Heseltine in 1973 who flew over Liverpool in a helicopter, I have been told, and said; 'that's where Wirral, Chester side starts, this is Wirral Merseyside, that's Sefton', and, you know, it was just where there was a bit of a gap in the population he defined the boroughs... [Knowsley] was kind of like everything they couldn't think of... it's like, what can we do with Huyton, then there's Kirkby at the other end, most of Cantril Farm, it's just weird". Thus, Knowsley borough, with its strange shape stretching around Liverpool almost like a protective shield, and its "unbalanced social structure" (Meegan, 1990, p92), appears to exist as the result of Heseltine's aerial carving up of the Liverpool region one afternoon in 1973.

5 These three bodies, all involving Heseltine to a greater or lesser extent, were essentially government-created organisations tasked with finding solutions to the problems of post-industrial Liverpool (Couch, 2003).
6 The 'Three Queens' is the marketized title created for the 175th anniversary of the creation of the Cunard cruise line, which saw the Queen Elizabeth, Queen Victoria, and the Queen Mary II visit Liverpool in May 2015 (Longman, 2016).
7 Enid refers here to St Luke's church on the corner of Leece Street and Berry Street, which was bombed during the Second World War and is now used as an open-air arts venue (Lees, 2013).
8 Both Mary and Jill referred to James Bulger using the more informal name Jamie; this is likely to be a reflection of the ongoing media fascination with the case and the media construct of 'Little Jamie Bulger' (Davis and Bourhill, 1997). Both of James' parents appear always to refer to him publicly as James, never Jamie.
9 The 'Gold Zone', established in 1998, was a public/private initiative created to facilitate capitalist prosperity in Liverpool city centre by 'sanitising' the streets of various perceived public incivilities; ranging from litter and graffiti to "unwanted or untoward activities" (Coleman, 2004, p215).
10 Teresa is referring here to an episode of *Boys from the Blackstuff*, 'George's Last Ride', in which George, an elderly, infirm man in a wheelchair, is pushed along Wapping by Chrissie, with the derelict and defunct Albert Dock pictured in the background (Millington and Nelson, 1986). Susie said of the drama: "*Boys from the Blackstuff*, now I think that should be compulsory viewing for everyone to see, and the desperation that people were living in, and going through".
11 The site re-opened as Festival Gardens in 1985, after which much of the land was sold off for housebuilding. The remaining land was used for several leisure purposes, including laser tag and paintballing, before falling into dereliction by 1997 (Wright, 2014).
12 051 was a 1990s Liverpool nightclub, created as a dance and rave venue to rival Manchester's Hacienda Club (Du Noyer, 2002).
13 The State Ballroom on Dale Street has variously been a traditional dance hall (Brocken, 2010), an alternative, 'indie' music venue (Lees, 2013), and part of the 'Scouse house' dance music scene (Du Noyer, 2002).
14 Ravetz (2001, p179) notes that "From time to time authorities tried to redeem a problem estate by an official change of name…Merseyside's Cantril Farm was, with tenure changes, given a new identity as Stockbridge Village". Roy said of the name change; "I've always been of the opinion that, let's show people what we've turned around rather than bury a name…I've got a real thing with renaming things just to make things seem better quickly, I'd rather say, look yeah, it's had a bad name in the past, let's move on with it".
15 *Brookside*, one of Channel 4's flagship programmes on its launch in 1982, was a soap opera set in Brookside Close on a 'new build' private estate in Liverpool. The programme ran for 21 years and is credited with bringing a new, authentic realism to the soap opera genre (Smith, 2007). The layout and 'look' of Brookside Close, synopsised by Grindrod (2013, p277) as "red brick…houses [with] small windows…which no longer 'let the outside in', giving instead a mean, fortress-like element", is reflected in, for example, the Eldonian Village and other inner area 'closes' built during the 1980s.
16 See, for example, Merseyside Civic Society's (2016) case for the redevelopment of the area's terraced housing stock. The authors compare the houses portrayed in long-running soap opera *Coronation Street* to those of Brookside Close, and find the latter to be lacking; "Those television soap operas, which place an emphasis on community life, are set in joined-up buildings such as those of Coronation Street and Albert Square. It seems to be significant that the [now defunct] Liverpool soap opera, *Brookside*, which had a very different emphasis, was set in a suburban cul-de-sac" (p80).
17 The Lodge Lane area has a large Islamic community, incorporating a significant Somali population.

18 The family in question were well known throughout inner Liverpool in the early 2000s for their criminal and anti-social behaviour, stories about them appearing in the *Liverpool Echo* and on local radio and television news frequently.
19 The Kensington Regeneration online archive cites a report published by the European Institute for Urban Affairs; "the comprehensive study – which runs to well over 100 pages – reveals that Kensington Regeneration has already achieved 29 of the 32 targets that were set at the start of the programme, with 22 of them being exceeded by a considerable margin".
20 The Kensington NDC was set up to run for ten years and was concluded in 2010.
21 Irene's view is borne out by the evidence provided by Professor Lewis Lesley (n.d.), an expert in transport planning, to the Edge Lane Public Enquiry; "Widening Edge Lane will merely transfer the problem further into the city centre, and stifle rather than encourage economic expansion in the city centre due to the land required for parking".
22 Irene is referring here to the New Mersey Shopping Park, close to Liverpool John Lennon Airport.
23 Coming Home Liverpool is a social enterprise run by Ronnie Hughes, which funds the restoration of empty houses and recoups the outlay from rent paid by tenants, who they find and match to properties. Rents are always 'fair' and calculated in line with the average wage in the Liverpool area.
24 Lark Lane runs between Aigburth Road and Sefton Park and is known for a variety of 'bohemian' shops, restaurants, and bars, and its Victorian architecture (Visit Liverpool, n.d.).
25 Mary refers here to the Yellow Duckmarine Tour, which was an amphibious wartime vehicle repurposed for tours around Liverpool's Albert Dock, until it sank mid-tour in 2013 (BBC, 2013).
26 Susie refers here to the green area of land that was on the site of Liverpool One.
27 The new Royal Hospital in Liverpool was due to open on its new site in 2017, but remains unfinished as I write (May 2020). The project is one of hundreds that fell through as a result of the collapse of multinational organisation Carillion who was awarded the contracts before going into liquidation in 2018, with a debt of £7 billion (Neate and Davies, 2020). The current estimated opening date for the new hospital is 2022 (Syal, 2020).
28 Joey and his family have lived in Brisbane, Australia for 15 years now and, in the summer of 2017 myself, my husband and my son attended Joey's son's wedding there, along with several other family members. We were stunned when one of the 'groomsmen', who was acting as Master of Ceremonies, warned the wedding guests to watch their valuables as they were now in the company of Liverpudlians.
29 The primary school where Mary teaches has a higher than average number of children for whom English is an additional language, children who have additional learning requirements, and children who are on the At Risk register.
30 This annual three-day meeting regularly attracts in the region of 150,000 racegoers who bring with them a revenue of between £10 and £30 million (Edwards, 2019).
31 Susie appears to be mixed up here; the Friday has always been Ladies Day, while the Thursday is currently billed as Liverpool Day – almost as if it is an attempt to restrict people from Liverpool to attending only on their designated day of the meeting.
32 Residents driving a car are required to prove who they are and where they live, with, for example, a driving license and a utility bill, before being allowed through barriers to park up and access their homes. The barriers, colloquially referred to as Checkpoint Charlie, are usually not too much of an inconvenience; however, there are occasions when overzealousness can result in tensions and bad decisions. An example of this occurred a few years ago, when my husband attempted to drive out of our cul-de-sac to collect our son from after school club, but was prevented from doing so by the yellow police van blocking the end of the road. The officer in the van informed my husband that he was not allowed to enter the 'sterile area'. As is the case in many relationships,

our marriage includes a division of labour, and talking to police officers falls squarely in my domain; my husband summoned me, and I remember saying to the officer: 'that's not a sterile area, it's Melling Road, and furthermore it's the route to after school club'. The matter was resolved when a woman officer instructed the first officer to move the van, and apologised for our inconvenience. I include this anecdote here as it is indicative of the ways in which ideas and notions of home are distorted in the context of cultural events, and the state of our homes, the enjoyment of our homes, and the rhythms of our home lives, are disrupted more and more by the increased commercialisation of events like the Grand National.

33 Working on the assumption that all traffic through Aintree train station will be people going to the races, pedestrians are corralled down a temporary path marked out with metal barriers which leads directly to the pedestrian crossing at the racecourse entrance. This adds a significant amount of distance to walk for people who, for example, need to turn right out of the station like Susie.

34 The current Cavern is a cellar venue across the road from the site of the original Cavern club famously associated with The Beatles (Du Noyer, 2002).

35 Gary is very aware of gender inequality and expressed regret about some of his actions as a younger man in the 1970s; "In them days you could actually, this is terrible from a feminist, you'd go, I'll take you home tonight, and you wouldn't see that girl all night, but she'd be waiting for you by the door, but that's just how it was coming from the 60s and into the 70s and it was that mentality, but looking back now it must have been horrendous…the arrogance of us boys then".

36 Tony refers here to what are colloquially referred to as the 'giants', giant-sized puppets first commissioned by Culture Liverpool in 2008, who have returned twice since then to commemorate the sinking of the Titanic and the outbreak of the First World War (Fraser, 2015).

37 Peter made a similar point: "Why don't they put money into opening old churches, or old buildings, and let the homeless in there?"

38 A 'local connection' can only be established if "individuals have lived in a location for six of the last twelve months, or have close family ties and/or an established work history in the relevant area" (Dwyer et al., 2014, p6).

39 The event in question was the Liverpool International Waterfront Festival, hosted by Liverpool City Council and held at the Cunard Building. The purpose of the event was to encourage collaboration between waterfront cities which also have a regeneration agenda.

40 This being said, Dave did note at the end of the questionnaire, "When filling out this questionnaire, I noticed myself becoming annoyed. Maybe I am more connected to Liverpool than I think!"

41 Jill takes issue with this assertion: "They say this about Liverpool people having more of an emotional attachment than others based on what evidence, that's what I would say; based on what evidence? Where is it? Provide me with actual evidence then I'll believe you, otherwise you're talking a load of bollocks".

42 Roy refers here to Liverpool City Council being a Labour stronghold in recent years, headed up by a Labour elected mayor, within the remit of a Labour regional mayor.

43 Thompson (2015, p458) essentially argues that false consciousness remains embedded in advanced capitalism, as the "culture industry's ability to drain the cognitive resources of individuals through increasingly trite forms of aesthetic experience, all point toward the institutionalization of false consciousness".

Bibliography

Allen C (2008) *Housing Market Renewal and Social Class Formation.* Routledge: Abingdon. doi:10.4324/9780203932742.

Anderson J (2015) 'Liverpool on a world stage' *Conference address Liverpool International Waterfront Forum*, 3–4 June 2015, Liverpool.

Andrews D and Turner S (2011) 'Is the pub still the hub? *International Journal of Contemporary Hospitality Management* 24 (4) pp 542–552. doi:10.1108/09596111211226815.

Avery P (2007) 'Born again: From dock cities to cities of culture' in Smith M (ed) *Tourism, Culture and Regeneration*. CAB International: Wallingford, pp 151–162. doi:10.1079/9781845931308.0000.

BBC News (2011) 'Liverpool's Everyman Theatre auctions seats', 28 July 2011. Available at http://www.bbc.co.uk/news/uk-england-merseyside-14328526. Accessed 8 April 2017.

BBC News (2013) 'Yellow duckmarine sinks in Albert Dock in Liverpool'. Available at https://www.bbc.co.uk/news/uk-england-merseyside-22922039. Accessed 12 May 2020.

Beckett A (2015) *Promised You a Miracle: Why 1980–1982 Made Modern Britain*. Penguin: London.

Boland P (2007) 'Unpacking the theory – Policy interface of local economic development: An analysis of Cardiff and Liverpool' *Urban Studies* 44 (5/6) pp 1019–1039. doi:10.1080/00420980701320736.

Bourdieu P (1985) 'The genesis of the concepts of habitus and field' *Sociocriticism* 2 pp 11–24.

Brocken M (2010) *Other Voices: Hidden Histories of Liverpool's Popular Music Scene* 1930s–1970s. Ashgate: London. doi:10.4324/9781315599151.

Byrne M (2016) 'Entrepreneurial urbanism after the crisis: Ireland's "Bad Bank" and the redevelopment of Dublin's Docklands' *Antipode* 48 (4) pp 899–918. doi:10.1111/anti.12231.

Coleman R (2004) *Reclaiming the Streets: Surveillance, Social Control and the City*. Willan: Cullompton.

Couch C (2003) *City of Change and Challenge: Urban Planning and Regeneration in Liverpool*. Ashgate: Aldershot. doi:10.4324/9781315197265.

Coyles D (2013) 'Reflections on Titanic Quarter: The cultural and material legacy of an historic Belfast brand' *The Journal of Architecture* 18 (3) pp 331–363. doi:10.1080/13602365.2013.804855.

Crick M (1997) *Michael Heseltine: A Biography*. Penguin: London.

Culture Liverpool (2017) *67–17: 50 Summers of Love*. Available at https://www.cultureliverpool.co.uk/summer-of-love/. Accessed 8 May 2017.

Davis H and Bourhill M (1997) '"Crisis": The demonization of children and young people' in Scraton P (ed) *'Childhood' in 'Crisis'?* UCL: London, pp 28–57.

Degen M and Garcia M (2012) 'The transformation of the "Barcelona Model": an analysis of culture, urban regeneration and governance' *International Journal of Urban and Regional Studies* 36 (5) pp 1022–1038. doi:10.1111/j.1468-2427.2012.01152.x.

Doucet B (2013) 'Variations of the entrepreneurial city: Goals, roles and visions in Rotterdam's Kop van Zuid and the Glasgow Harbour megaprojects' *International Journal of Urban and Regional Research* 37 (6) pp 2035–2051. doi:10.1111/j.1468-2427.2012.01182.x.

Du Noyer P (2002) *Liverpool: Wondrous Place: Music from the Cavern to the Coral*. Virgin Books: London.

Dwyer P, Bowpitt G, Sundin E and Weinstein M (2014) 'Rights, responsibilities and refusals: Homelessness policy and the exclusion of single homeless people with complex needs' *Critical Social Policy* 35 (1) pp 3–23. doi:10.1177/0261018314546311.

Edwards L (2019) 'How the Grand National benefits the local economy' *Click Liverpool*, 28 March 2019. Available at https://www.clickliverpool.com/features/32370-how-the-grand-national-benefits-the-local-economy/. Accessed 12 May 2020.

Ferreira S and Visser G (2007) 'Creating an African Riviera: Revisiting the impact of the Victoria and Alfred waterfront development in Cape Town' *Urban Forum* 18 (3) pp 227–246. doi:10.1007/s12132-007-9008-3.

Fraser I (2015) 'Mayor Joe Anderson hints at Giants return to Liverpool in 2016' *Liverpool Echo*, 21 January 2015. Available at http://www.liverpoolecho.co.uk/news/liverpool-news/mayor-joe-anderson-hints-giants-8424325. Accessed 21 May 2017.

Foucault M, Davidson A and Birchall G (2008) *The Birth of Biopolitics: Lectures at the College de France 1978–79*. Palgrave: Basingstoke.

Frost D and North P (2013) *Militant Liverpool: A City on the Edge*. Liverpool University Press: Liverpool. doi:10.5949/liverpool/9781846318634.001.0001.

Furmedge P (2008) 'The regeneration professionals' in Allt N (ed) *The Culture of Capital*. Liverpool University Press: Liverpool.

Garland D (2001) *The Culture of Control: Crime and Social Order in Contemporary Society*. Oxford University Press: Oxford. doi:10.7208/chicago/9780226190174.001.0001.

Grindrod J (2013) *Concretopia: A Journey around the Rebuilding of Postwar Britain*. Old Street Publishing: Brecon.

Heseltine M and Leahy T (2011) *Rebalancing Britain: Policy or Slogan? Liverpool City Region – Building on Its Strengths: An Independent Report*. Department for Business, Education and Skills: London.

Hollander J (1991) 'It all depends' *Social Research* 58 (1) pp 31–49.

Kensington Regeneration (n.d.) *Kensington Regeneration Hailed as a Success*. Available at http://www.kensingtonregeneration.org/news_and_events/press/kensington_regeneration_hailed_as_a_success/index.html. Accessed 9th April 2017.

Kensington Regeneration (2008) *Deliverington: Kensington Regeneration Delivery Plan 2008/9*. Available at http://kensingtonregeneration.org/images/uploads/delivery_plan.pdf. Accessed 2 June 2017.

Kidd S and Evans D (2011) 'Home is where you draw strength and rest: The meanings of home for houseless young people' *Youth & Society* 43 pp 752–773. doi:10.1177%2F0044118X10374018.

Lees A (2013) *Liverpool: The Hurricane Port*. Mainstream Publishing: Edinburgh.

Lesley L (n.d.) *Edge Lane Public Inquiry: Evidence from Professor Lewis Lesley*. Available at www.edgelaneliverpool.co.uk/download/i/mark_dl/u/.../Prof_LJL__transport_.pdf. Accessed 9th April 2017.

Liverpool City Council (2016) *Better Roads City Centre Connectivity Scheme*. Available at https://liverpool.gov.uk/parking-travel-and-roads/better-roads/better-roads-schemes/completed-schemes/city-centre-connectivity-scheme/. Accessed 6 May 2017.

Longman D (2016) *Liverpool in the Headlines*. Amberley: Stroud.

Meegan R (1990) 'Merseyside in crisis and conflict' in Harlow E, Pickavance C and Urry J (eds) *Place, Policy and Politics: Do Localities Matter?* Unwin Hyman: London, pp 87–107.

Merseyside Civic Society (2016) *Terraced Housing in the Liverpool City Region: The Most Sustainable and Attractive Option*. Available at http://www.merseysidecivicsociety.org/media/78583/wp2-terraced-housing-in-the-liverpool-city-region-v1.pdf. Accessed 12 May 2020.

Millington B and Nelson R (1986) *'Boys from the Blackstuff': The Making of TV Drama*. Comedia: London.

Neate R and Davies R (2020) 'Carillion collapse: Two years on, "government has learned nothing"' *The Guardian*, 15 January 2020. Available at https://www.theguardian.com/business/2020/jan/15/carillion-collapse-two-years-on-government-has-learned-nothing. Accessed 12 May 2020.

Prior J and Blessi G (2012) 'Social capital, local communities and culture-led urban regeneration processes: The Sydney Olympic park experience' *Cosmopolitan Civil Societies: An Interdisciplinary Journal* 4 (3) pp 78–96. doi:10.5130/ccs.v4i3.2684.

Ravetz A (2001) *Council Housing and Culture: The History of a Social Experiment*. Routledge: London. doi:10.4324/9780203451601.

Roberts L (2012) *Film, Mobility and Urban Space: A Cinematic Geography of Liverpool*. Liverpool University Press: Liverpool. doi:10.5949/upo9781846317248.

Scraton P (2009) *Hillsborough: The Truth*, 20th anniversary ed. Mainstream Publishing: Edinburgh.

Skeggs B (2004) *Class, Self, Culture*. Routledge: London. doi:10.4324/9781315016177.

Smith P (2007) '"I've got a theory about Scousers": Jimmy McGovern and Lynda La Plante' in Jones D and Murphy M (eds) *Writing Liverpool: Essays and Interviews*. Liverpool University Press: Liverpool, pp 210–227. doi:10.5949/UPO9781846314476.016.

Syal R (2020) 'Two hospitals held up by Carillion collapse are delayed further' *The Guardian*, 17 January 2020. Available at https://www.theguardian.com/society/2020/jan/17/two-hospitals-held-up-by-carillion-collapse. Accessed 12 May 2020.

Theokas A (2004) *Grounds for Review: The Garden Festival in Urban Planning and Design*. Liverpool University Press: Liverpool. doi:10.5949/upo9781846314643.

Thompson M (2015) 'False consciousness reconsidered: A theory of defective social cognition' *Critical Sociology* 41 (3) pp 449–461. doi:10.1177%2F0896920514528817.

Tulloch A (2011) *The Story of Liverpool*. The History Press: Stroud.

Varna G (2014) *Measuring Public Space: The Star Model*. Ashgate: London. doi:10.4324/9781315594408.

Visit Liverpool (nd) *Lark Lane*. Available at Click here to enter text.http://www.visitliverpool.com/explore-the-city/top-spots/lark-lane. Accessed 10th April 2017.

Wilson H and Womersley J (1977) *Change or decay: Final report of the Liverpool Inner Area study*. HMSO: London.

Wright J (2014) 'Nostalgia: Remembering when Liverpool's International Garden Festival opened its gates' *Liverpool Echo*, 2 May 2014. Available at http://www.liverpoolecho.co.uk/news/liverpool-news/nostalgia-remembering-liverpools-international-garden-7066782. Accessed 6 April 2017.

8 Pierre Bourdieu, *habitus*, and the ‘new’ Liverpool home

Data and analysis presented in Chapters Six and Seven point clearly to an important conclusion: That the concepts of home and regeneration in the Liverpool context are different depending on the time period under consideration. Respondents repeatedly made connections between the past and their immediate dwelling and locale, whilst linking the present to the city centre, the waterfront, and the city’s global standing. Arguably, the difference between past understandings of regeneration and home, and present understandings of regeneration and home, constitutes a significant finding in relation to respondents’ understandings and conceptualisations of the Liverpool home. To reiterate, respondents almost universally connect their Liverpool home of the past with the district that they grew up in, whilst their contemporary Liverpool home is connected with the city centre.

When asked to describe the home of their childhood, participants discussed their early dwellings to a point, but spoke much more, and much more fondly, about the surroundings of the dwelling and the relationships that characterised them. Extended family, quasi-extended family in the form of neighbours, patterns of play, and neighbourhood rituals and rhythms were recalled as signifiers of home, evoking familiarity, comfort, and nostalgia. What is very clear from the data is the sense that, when respondents are asked to recall their memories of the city dating back to the ‘forward-facing’ era of the 1950s to the early 1980s, they almost universally interpret this request as pertaining to the buildings within which they lived, their family and friendship networks, and their immediate local areas and districts. Those who *did* discuss, for example, the city centre, did so either (a) because they were specifically prompted to do so, (b) they lived close to the city centre, or worked within it – thereby rendering it more central to their memories, or (c) they mentioned it briefly in the abstract.

However, when asked to consider Liverpool and home in the contemporary context, respondents focused squarely on the city centre and, in particular, the waterfront. This position differs significantly from the respondents’ thoughts, ideas, and memories pertaining to the ‘forward-facing’ era as it almost universally focused on developments in the city centre, or developments concerning Liverpool’s place in a global setting. There were some exceptions, for example, the seemingly bungled regeneration of Kensington, and the organic renaissance of Lodge Lane; but these were supplemental to the overarching narrative of the city centre as emblematic of the contemporary regenerated Liverpool home. So, from the data, it is possible to present respondents' conceptualisations of Liverpool and home as simple formulae:

Home + Liverpool + the past = Liverpool's districts and neighbourhoods.
Home + Liverpool + the present = Liverpool's city centre and waterfront.

This chapter, then, considers this distinction, possible reasons for its occurrence, and posits that the work of Pierre Bourdieu (1985) on the concepts of *habitus* and *field* might provide a useful mechanism for understanding the gradual drift from one way of thinking about regeneration, home, and Liverpool to another one.

In terms of perceptions of regeneration and home in Liverpool during the forward-facing period, turning to established ways of thinking in relation to home and the passage of time is helpful. It is commonplace for people to feel a sense of nostalgia for the past and, consequently, to focus on the aspects of the past which are more personal, more intimate, and more indicative of a sense of belonging (Morley, 2000; Waghorn, 2009; Lewicka, 2010; Mac an Ghaill and Hayward, 2011). Therefore, it can be argued that the midpoint between dwelling or residence and home town or home city – one's neighbourhood, district, or *baille* – provides people with greater sense of belonging, "experience of locality" (Galbraith, 1997, p122), and shared characteristics with others than the immediate home as in household, and simultaneously offers a greater sense of intimacy, affinity, concord, and camaraderie than the more 'macro' level home city. In this sense, home as community, *mahalla*, or *marae* is more evocative of the past, and one's sense of identity and, crucially, the person one *used to be*. Just as *heimweh* is more than homesickness, in that it is a longing for something that can never be regained in the same way, recollections of home may be read to extend no further than the neighbourhood, district, or locality which holds the most symbolic meaning; sensations of home appear to be lost beyond one's immediate locale with the passage of time.

A further explanation that must be considered is the nature of the changes to home in terms of regeneration processes that occurred during the forward-facing period. Whilst some remodelling and modernisation did occur in the Liverpool city centre between the 1950s and the start of the 1980s, arguably, it was minimal in comparison to the wholesale changes we have seen from the early 1980s onwards. Regeneration and development of the *districts*, however, was in many cases extensive, with some neighbourhoods in Everton, Kirkdale, and Edge Hill, for example, disappearing altogether, and others like parts of Kirkby, Cantril Farm, and Netherley built up from scratch. In this sense, it is perhaps unsurprising that people will recall changes to their immediate environment much more than changes in a city centre that they may have only visited infrequently.

Another reason for prioritisation in the imagination of respondents of neighbourhood over dwelling at one end of the spectrum, and home city at the other, may simply be the driving force behind the regeneration – the reasons why it was deemed necessary, and the image and message that those controlling the city wanted to establish. Housing need was perceived to be the chief motivation behind the restructuring of the districts (Muchnick, 1970) and, as demonstrated in the literature, changing fashions and social mores dictated that 'modern' housing should shun the city centre, should embrace progressive and experimental architecture, should explore street layouts more conducive to modern life, and, crucially, should facilitate social engineering and the furtherance of capitalist systems (Grindrod, 2013). The focus being entirely on the future meant that culture, heritage, and history played no role in the planning of regeneration and development – processes of regeneration were predominantly about housing, and the past had no place in the houses of the future. Thus, the city centre would, by its very nature, register less in the memory of people asked to think about home, regeneration, and Liverpool in the 1950s, 1960s, and 1970s.

It is therefore a justifiable conclusion to draw that the nature of nostalgia and memory, and the dual focus of the 'forward-facing' period on both housing and the modernisation/development of districts, converge here with the result of 'home' being conceptualised in memories of this era at the micro, or at most the intermediate, level. Conversely, conceptualisations of the Liverpool 'home' during the contemporary 'backward-facing' period focus significantly more on the city centre. This association between the past and the districts, and the present and the city centre, may simply be a reflection of the processes of growing up and becoming an adult and, significantly, the places and spaces associated with these processes. Children growing up in cities tend to gain freedom of movement from their parents incrementally, first accessing their immediate environs, then gradually exploring further away from home, until they gain full, entitled access to the city centre, through employment, shopping, and leisure. In this sense alone, the city centre may be emblematic of personal freedom and independence. However, the notion that the districts represent the home of the past, and the city centre represents the home of the present, is at odds with the historical and geographical development of the city, as the city was 'born' from the centre and grew outwards. The picture is further complicated by the contemporary 'regenerating' city centre focus on the past, preoccupied with commemorating and festivalising past events. Thus, a sense of home in the city centre is created; a conceptually strange hybrid of the past and the present, with hopes for the future more so than certainties.

Participants' apparent contemporary alignment of home as a concept with the city centre is arguably a product of the pride that was frequently conveyed during the interviews, and the evident pleasure at Liverpool being under the media, cultural, and political spotlight for something good for a change: For important events and processes of the past which have been heritagised and repackaged as evidence of a brighter today and an even brighter tomorrow. Given that home in all its forms is conceptualised as an extension of the self (Kearon and Leach, 2000; Fox, 2002), if Liverpool is prospering, all those for whom Liverpool is home are also prospering. Instead of being globally vilified for football violence, social unrest, entrenched worklessness, political militancy, or whatever other pejorative judgements that have been made on the city and its populace, Liverpool is now constructed as a 'global city', claiming its place on the 'global stage', the envy of other cities that do not have our river, our historic built environment, our people, or our culture. Global status has been achieved through the statecraft-sanctioned management of the city centre to present the city's best self to the wider world, including those who enhance the regeneration narrative and excluding those who undermine it. To borrow again from respondent Mary's turn of phrase, in this sense the city centre is the proverbial 'fur coat': Impressive, and adept at obscuring the 'no knickers' nature of the districts – some doing well, some doing less well, but all outside of the contemporary regenerational and cultural spotlight.

Habitus and Home

Chapter Two established the conceptual framework underpinning the primary research and the project as a whole – the idea of home, how it developed as a concept, what it means, the levels upon which it operates, and how it can be understood to be a *process* as much as a *place*. The relationship between home and time, and the complexity of home as at once public and private, freedom and restriction, and belonging and rejection were explored, together with the role that home as a concept plays in the maintenance and furtherance of advanced capitalism. To better understand and crystallise home, its polar opposite – homelessness – was conceptualised, alongside processes which lead to loss of home up to and including the deliberate destruction of home – domicide. This section takes home and its associated meanings and concepts and examines them in relation to Bourdieu's (1977, 1983, 1985) understanding of *habitus*; "the ways in which the social is literally incorporated" within individuals and groups (Lawler, 2004, p111).

As outlined in Chapter One and expanded upon in Chapter Two, *habitus* is a Bourdieusian concept employed to signify dispositions or socialised norms which are controlled, subtly and inconspicuously, by social and political elites who have the power to guide, shape, and ultimately control broad-brush views, opinions, and perceptions (Bourdieu, 1983; Jenkins, 2002; Lawler, 2004). *Habitus* is the means by which we are *taught* what we should *know*, as prescribed by the social and political elite, without even realising we are being schooled:

> The habitus is the product of inculcation and appropriation necessary in order for those products of collective history, the objective structures (for example, language, economy, and so on) to succeed in reproducing themselves.
>
> (Bourdieu, 1977, p72)

This occurs within an arena with clearly identified parameters and boundaries or, to use Bourdieusian terminology, a given *field*. Bourdieu (1977, p79) notes that "we are very much aware of the most recent attainments of civilisation, because, being recent, they have not yet had time to settle into our unconscious" – thus, over time, populations can be brought, via habitus, to a perspective wholly different to their previously held viewpoint, without even realising that they have undergone a shift in outlook, much less that they have been manipulated into such a shift via external forces (Elias, 2000). This is largely due to the time period – often lengthy – which is needed for the process of change to be concluded; habitus takes so long to 'work' that it is essentially "embodied history, internalised as second nature and so forgotten as history" (Bourdieu, 1990, cited in Lawler, 2004, p112).

It is in this way that habitus can provide an explanation for the chief finding of this project – that the Liverpool people sampled here equate regeneration, home, and the *past* with their immediate home and surrounding area, but clearly equate regeneration, home, and the *present* with Liverpool's city centre, the city at large, and the city as it is perceived by the wider world. One of the common ways that home as a concept is theorised is via the notion of 'levels' of home, from the micro scale (the bed, the hearth) to the macro scale (nation), and incorporating a range of 'layers' of home in between (dwelling, community, city, and so on); Haywood (1977) employs the metaphor of concentric rings radiating outwards from a central point. Data indicates that the regenerated home as a concept was attributed by the research sample very much to the micro end of the spectrum in respect to the past: Residence, family, street, church, school, and locality. At the same time though, data also points to the contemporary regenerated home being clearly conceptualised as much further up the scale, at city/city region level.

Regeneration, then, both in terms of actual change to built environments, and in terms of regeneration discourse, has arguably shifted priority from wholesale modern housing provision in the periphery during the forward-facing period, to both physical and reputational regeneration of the city centre, the waterfront, and the city's image as perceived by the wider world during the backward-facing period. Ravetz (2001) notes that the case of Liverpool is, regarding the provision of social housing, distinctive from patterns evident in other cities; forward-facing regeneration assumed that the existing trajectory of economic growth, requiring an ever-increasing workforce, would necessitate ever greater provision of dormitory accommodation at the outskirts. Instead, however, growth slowed down, peripheral housing provision was rendered virtually redundant and slipped into disrepair and neglect as focus returned to the city centre. Arguably, a gradual repositioning of habitus, which nudges societal understandings of what home means in Liverpool away from the immediate home and district and towards the city centre and the city's global standing, has facilitated greater acceptance on the part of Liverpool people of re-prioritisation of the city centre at the expense of the districts.

In this way, the meaning of the regenerated home in Liverpool, and the way it is experienced and understood, has evolved from the dwelling/neighbourhood level to the city level, without the change being noted or even noticed. Although home is an exceptionally complex, multi-layered, nebulous, and subjective concept, it appears at first sight as if it is simple and one-dimensional; its ubiquity and universality render it mundane and workaday, obscuring its convolution (Rybczynski, 1986). Because of this, in the normal way of things, home as a concept goes unchallenged – it tends not to be interrogated as a notion due to its 'taken for granted' nature (Hollander, 1991). It is reasonable to speculate that this lack of probing enhances and feeds into the *habitus* which leads Liverpool people to accept the contemporary construction, for regeneration and other purposes, of Liverpool city centre as the embodiment of home rather than its individual districts, estates, streets, and dwellings. Mallet (2004) contends that home is a process more so than it is a location or a building, it is something that we *do* rather than something that we *have*; group and individual *habitus* can be seen as having gradually altered our 'doing' of home to be as much about, for example, attending mega-events at the waterfront, as domestic activities such as laundry, gardening, and home maintenance. By engaging with such events, we reinforce the notion of the Liverpool home as operational at the city rather than district level.

One of the ways in which this is being achieved is via nostalgia, memories of and affection for previous times. When something has a history and a heritage, it is afforded a certain gravitas; a level of dignity, respect, and reverence that cannot be attributed in the same way to other phenomena. Because Liverpool as a city has an abundance of history and heritage, feelings of importance, significance, and pride facilitate a deep and widely held love for those aspects of the city revered as valuable, distinctive, and evidential of a glorious past. Arguably, if the Titanic had not sunk on its maiden voyage, the UK had not joined the Common Market in 1973, and The Beatles had remained a local band working the pub circuit, the past as we currently understand it in Liverpool would be the same as the present, unworthy of the level of sentimental pride and commercial value it is now attributed. In this sense *heimweh* – homesickness for a circumstance, or way of being, that can never be regained (Hollander, 1991) – is harnessed to gain unwitting acceptance of, and support for, the contemporary event-led regeneration which holds history and heritage as its most valuable assets.

Because of its geography, Liverpool has always had a sense of transience about it, as its natives went to sea, visitors passed through *en route* elsewhere, and immigrants made what they hoped was a temporary home before moving on to somewhere they believed would offer more opportunity (Christian, 1997; Belchem and MacRaild, 2006; Aughton, 2008; Balderstone et al., 2014). Thus, transience and mobility are woven into the fabric of the city, as is stasis for those whose mobility is hampered. Value is attached to the mobility, both physical and social, of the "Liverpolitan diaspora" (Heseltine and Leahy, 2011, p43), and the Liverpool-specific habitus which encourages acceptance of contemporary conceptions of the Liverpool home as the waterfront is far-reaching. The majority of 'ex-pat' Liverpudlians who engaged with my project fully embraced the notion of a rehabilitated Liverpool with much to offer the world, and were both supportive of, and comfortable with, the notion of themselves as ambassadors for the city. Even if their understanding of contemporary Liverpool is "an imaginative construct", as suggested by one of my respondents (Jill), its potency is enough to foster within them nostalgia, civic pride, and support for a corporate and commercial renaissance.

It must also be noted that home, whether it be viewed as a place, a space, or a process, has been harnessed on all its levels for its ability to promote and maintain capitalist agendas. Both the home furnishing industry and the home 'lifestyle' industry – incorporating everything from dinner parties to home spa treatments to Easter/Father's Day/Halloween celebrations – continue to generate huge amounts of money (John Lewis Communications, 2019). Academics have cautioned against the conflation of home with financial asset (Hollander, 1991; Mallett, 2004; Coolen and Meesters, 2012), yet all aspects of home are permeated by the continual push towards capitalist accumulation. As both a showcase of the self and an indicator of status and personal wealth (Cuba and Hummon, 1993; Cristoforetti et al., 2011), home can be viewed as evidence that one has taste, style, and class; equally, it can disclose lack thereof. Constructions of home at the *city level* as cultured, prosperous, and distinctive act to support these existing views and processes intrinsically linking home in the Liverpool context to broader capitalist mechanisms and drives. *Fremde* – or, the sensation of being unwanted and excluded in a foreign, unfamiliar place (Morley, 2000) – becomes the conceptual domain of those unable to embody such advanced capitalist renditions of home. 'Street life' people, such as rough sleepers, street drinkers, and beggars, are allowed neither space nor place within this cultured, highly capitalist setting.

Heim and *heimat*, the German terms used to distinguish between the private, small-scale, immediate home, and the public home in the sense of nationality and 'fatherland', provide us with a useful starting point in further unpacking and interrogating the concept. However, what appears to be evident is that home cannot simply be reduced to a straightforward binary of public and private; rather, a taxonomy of home as being multi-layered, as indicated by Haywood (1977), is more useful in reflecting and theorising the multiple forms that home takes. My findings give rise to a few important ideas. First, that the notions of 'private' and 'public' in relation to home are becoming increasingly blurred, in that processes including the advent and use of social media and the growing move towards entertaining at home render the dwelling less private. Similarly, in the Liverpool case at least, stakeholder mindsets driven by sentimentality, nostalgia, and pride, together with more and more space becoming quasi-public/quasi-private, mean that the public nature of home, as in home city, is effectively made private. Second, and relatedly, financial ownership of a dwelling or dwellings is not a good indicator of sensations and feelings of home; people 'feel' ownership of home, be it their property or otherwise, that bears no relation to ownership in the legal sense.

Third, as established earlier, the tendency amongst my respondents was to link regeneration and home in the past to the smaller end of the scale if, for example, employing Haywood's (1977) model of concentric rings – immediate home, family, neighbourhood – whilst linking home, regeneration, and the present to Liverpool's city centre and waterfront. A case can be made that this has facilitated, via *habitus*, acceptance that regeneration is something that happens in the city centre, and that revival in the districts will either not materialise, or will arise from community activity rather than being funded and 'delivered' by Liverpool City Council or other agencies. In a manner reminiscent of the land reclamation and urban farming that has emerged out of the dereliction of Detroit, as identified by Julian Temple in *Requiem for Detroit?* (2010), Walton, a district to the north of the city centre, is one such area where, frustrated with lack of action from the local authority, local people have come together in an attempt to rejuvenate and regenerate the area. To date, this has included local business owners investing further in the area, 'guerrilla' gardening on sites left vacant since the Second World War, street art, alleyway adoption, and volunteer street cleaning (County Road Regeneration, 2017). In this sense, the residents of Walton are engaging in resistance to the continual push towards centralisation in terms of conceptions of home in Liverpool. Arguably, the academy should do more to conceptualise and theorise home, its various meanings, and associated implications for urban regeneration, at this community/neighbourhood/district level – "the midpoint of the scale of place continuum" (Lewicka, 2010, p36). The binary conceptualisation of *heim* – one's dwelling, and *heimat* – one's homeland, is too narrow and simplistic to explain the importance attributed to one's immediate locality, particularly in relation to the past homes of our memories.

Bibliography

Aughton P (2008) *Liverpool: A People's History*, 3rd ed. Carnegie Publishing: Lancaster.

Balderstone L, Milne G and Mulhearn R (2014) 'Memory and place on the Liverpool waterfront in the mid-twentieth century' *Urban History* 41 (3) pp 478–496. doi:10.1017/S0963926813000734.

Belchem J and MacRaild D (2006) 'Cosmopolitan Liverpool' in Belchem J (ed) *Liverpool 800: Culture, Character and History*. Liverpool University Press: Liverpool, pp 311–391.

Bourdieu P (1977) *Outline of a Theory of Practice*. Cambridge University Press: Cambridge.

Bourdieu P (1983) 'The field of cultural production, or: The economic world reversed' in Johnson R (ed) (1993) *The Field of Cultural Production: Essays on Art and Literature*. Polity: Cambridge, pp 29–73.

Bourdieu P (1985) 'The genesis of the concepts of habitus and field' *Sociocriticism* 2 pp 11–24.

Christian M (1997) 'Black identity in Liverpool: An appraisal' in Ackah W and Christian M (eds) *Black Organisation and Identity in Liverpool: A Local, National and Global Perspective*. Charles Wootton College Press: Liverpool, pp 62–79.

Coolen H and Meesters J (2012) 'Editorial special issue: House, home and dwelling' *Journal of Housing and the Build Environment* 27 pp 1–10. doi:10.1007/s10901-011-9247-4.

County Road Regeneration (2017) Available at https://www.facebook.com/groups/414832075281953/. Accessed 19 September 2017.

Cristoforetti A, Gennai F and Rodeschini G (2011) 'Home sweet home: The emotional construction of places' *Journal of Aging Studies* 25 pp 225–232. doi:10.1016/j.jaging.2011.03.006.

Cuba L and Hummon D (1993) 'A place to call home: Identification with dwelling, community and region' *The Sociological Quarterly* 34 (1) pp 111–131.

Elias N (2000) *The Civilising Process: Sociogenetic and Psychogenetic Investigations*. Blackwell: Oxford.

Fox L (2002) 'The meaning of home: A chimerical concept or a legal challenge?' *Journal of Law and Society* 29 (4) pp 580–610. doi:10.1111/1467-6478.00234.

Galbraith M (1997) '"A Pole can die for the fatherland, but can't live for her": Democratisation and the Polish heroic ideal' *The Anthropology of East Europe Review* 15 (2) pp 119–139.

Grindrod J (2013) *Concretopia: A Journey around the Rebuilding of Postwar Britain.* Old Street Publishing: Brecon.

Hayward G (1977) 'Psychological concepts of home' *HUD Challenge* 8 (2) pp 10–13.

Heseltine M and Leahy T (2011) *Rebalancing Britain: Policy or Slogan? Liverpool City Region – Building on Its Strengths: An Independent Report.* Department for Business, Education and Skills: London.

Hollander J (1991) 'It all depends' *Social Research* 58 (1) pp 31–49.

Jenkins R (2002) *Pierre Bourdieu,* 2nd ed. Routledge: London.

John Lewis Communications (2019) *The John Lewis Retail Report 2019: How We Shop, Live and Look.* Available at https://www.johnlewispartnership.co.uk/content/dam/cws/pdfs/media/how-we-shop-live-and-look-2019.pdf. Accessed 15 May 2020.

Kearon T and Leach R (2000) 'Invasion of the "body snatchers": Burglary reconsidered' *Theoretical Criminology* 4 (4) pp 451–472. doi:10.1177%2F1362480600004004003.

Lawler S (2004) 'Rules of engagement: Habitus, power and resistance' *Sociological Review* 52 (2) pp 110–128. doi:10.1111%2Fj.1467-954X.2005.00527.x.

Lewicka M (2010) 'What makes a neighbourhood different from home and city? Effects of place scale on place attachment' *Journal of Environmental Psychology* 30 pp 35–51. doi:10.1016/j.jenvp.2009.05.004.

Mac an Ghaill M and Haywood C (2011) '"Nothing to write home about": Troubling concepts of home, racialization and self in theories of Irish male (e)migration' *Cultural Sociology* 5 (3) pp 385–402. doi:10.1177%2F1749975510378196.

Mallett S (2004) 'Understanding home: A critical review of the literature' *Sociological Review* 52 (1) pp 62–89. doi:10.1111/j.1467-954X.2004.00442.x.

Morley D (2000) *Home Territories: Media, Mobility and Identity.* Routledge: Abingdon. doi:10.4324/9780203444177.

Muchnick D (1970) *Urban Renewal in Liverpool.* The Social Administration Research Trust: Birkenhead.

Ravetz A (2001) *Council Housing and Culture: The History of a Social Experiment.* Routledge: London. doi:10.4324/9780203451601.

Requiem for Detroit? (2010) TV, BBC2, 13 March 21.00.

Rybczynski W (1986) *Home: A Short History of an Idea.* Pocket Books: London.

Waghorn K (2009) 'Home invasion' *Home Cultures* 6 (3) pp 261–286. doi:10.2752/174063109x12462745321507.

9 Conclusion

The Lego Movie, released to critical and popular acclaim in 2014, depicts a world of uniformity, conformity, and homogeneity where following the instructions is vital in order "to fit in, have everyone like you, and always be happy". Bricksberg, a city controlled jointly by President Business and mega-corporation Octan, is a place where everyone supports the same local sports team, has the same favourite television programme (*Honey, Where are my Pants?*), the same favourite song ("Everything is Awesome"), and the same favourite restaurant ("any chain restaurant"). Take out coffee is expensive, Tuesdays are always tacos for tea, and road signs instruct motorists and pedestrians to "integrate". The film's protagonist, Emmet, is a 'generic' construction worker who wants to fit in yet has a vague sense that something is wrong with the ordered nature of everyday life. He encounters a rogue group of "master builders" who refuse to conform to the "instructions" and seek to create a lifestyle and a space where individuality and creativity are valued; a place where "relics" from the time before the reign of President Business/Octan are respected for their quirky, unique qualities. The master builders are outlawed by President Business, constructed as selfish and self-interested and a threat to the uniformity, and therefore the prosperity, of the city.

The Lego Movie can be read as a metaphor for the contemporary, prosperous, consumerist city. Construction work is ubiquitous and continual, shops, hotels, and restaurants are multi-national and common to other prosperous cities, and government is a joint venture between the public and the private. Importantly, there is consensus amongst the populace that all is well – prosperity is good for everybody; therefore, everybody must support the neoliberal, statecraft, conformist agenda. Those who question or challenge this prosperity agenda are constructed as outsiders, disloyal to the common cause, and unworthy of belonging and the benefits that it brings. Liverpool is now one such city, bearing all the hallmarks of a contemporary European 'city break' destination. Liverpool is constructed as on an upward trajectory, reliant, not on industry or trade, but other resources – heritage, culture, nightlife, shopping, and a friendly, welcoming, entertaining populace who are 'on message' with Liverpool's current direction. Just as, for example, Cantril Farm was renamed as Stockbridge Village alongside other reforms driven by Michael Heseltine (Crick, 1997), Liverpudlians who are willing to collude with contemporary processes are seen to have evolved into 'Liverpolitans' – a term free of the negative baggage of the Liverpool of the 1970s and 1980s.

In completing this project, I set out to find answers to two key research questions:

1. What impact has the 'forward-facing' regeneration of the 1950s–1970s had on experiences of and feelings about the Liverpool home?
2. What impact has the 'backward-facing' regeneration of the 1980s onwards had on experiences of and feelings about the Liverpool home?

Derived from my interpretation and understanding of the literature on Liverpool in a regeneration context, the key research questions facilitated data collection and analysis which reflect the stark differences in approach, attitude, and mindset in terms of how regeneration could, or should, be achieved. Question One reflects the mood or climate of the post-war era, characterised by the assumption that Liverpool would continue to be a prosperous port; Question Two, on the other hand, reflects the slow acceptance that previous paths to prosperity were no longer open, and other, more nebulous regeneration initiatives must be developed.

The narrative themes, derived from the grounded theory method and developed in the analysis chapters, have been distilled into clear, straightforward responses to the key research questions. To reiterate once more, thinking about the 'forward-facing' period of regeneration in relation to home and Liverpool led respondents to discuss their Liverpool home from that period as their immediate dwelling and, perhaps more so, their local neighbourhood. Conversely, thinking about the 'backward-facing' period of regeneration, with its focus on heritage, history, and culture, led respondents to identify their more recent or contemporary understandings of home in Liverpool as related to the city centre; more specifically, the waterfront. The regenerated Liverpool home of the post-war era is the neighbourhood; the regenerated Liverpool home of the contemporary era is the city centre.

Thus, a significant shift has occurred, which appears to mirror the change in focus apparent in the literature, where regeneration of the city morphed from the physical and peripheral – 'slum' clearance, house building, and road restructuring in the districts – to the reputational and central; harnessing the historic nature of the city centre and the waterfront to establish cultural regeneration as the route out of trouble and the path to prosperity. The nature of this shift, from demolition of the existing built environment, and construction on a blank canvas, to restoration, preservation, and celebration of the historical detritus of a maritime, industrial past, has arguably caused my respondents to shift their understanding of their Liverpool home, from the micro – dwelling, extended family, local area, to the macro – the city centre, the waterfront, and the city as it is viewed in the wider world.

My contention is that this gradual shift in conceptualisations of one's 'Liverpool home' may be understood and made sense of by employing Bourdieu's (1977, 1983, 1985) concepts of *habitus* and *field*. Bourdieu holds that the *habitus* is essentially an unconscious following of rules, norms, and social mores which become "second nature" (1968, p234), so much so that individuals are unaware even of the existence of the rules that they follow, much less that they conform to them: "the power of the *habitus* derives from the *thoughtlessness* of habit and habitation, rather than consciously learned rules and principles" (Jenkins, 2002, p76; emphasis added). While *habitus* occurs on an individual level, it also functions on the level of social class, whereby a working-class *habitus* becomes identifiable via "analogous preferences across a broad range of cultural practices" (Johnson, 1993, p5).

The *fields* meanwhile are the contexts within which the *habitus* operates and are categorised according to their characteristics; the political *field*, the educational *field*, and so on (Bourdieu, 1985). While each *field* has its own laws of functioning, and its own hierarchies, *fields* are all related to each other, and fit together with each other like children's building blocks. I posit that within a neoliberal *field* characterised by regeneration, culture, and civic pride, *habitus* could be the process by which people from Liverpool are led to unwittingly shift their perceptions of a regenerated 'Liverpool home' from dwelling and immediate neighbourhood to city centre, and Liverpool's place on the global stage. Over a period of 30-plus years, people for whom Liverpool is home have slowly and gradually been brought to the view that the city's cultural and maritime heritage is its greatest asset, so gently and insidiously that the change of view was almost imperceptible.

This position is further supported by Bourdieu's linking of *habitus* and *field* to both history and culture. Jenkins (2002, p80) notes that "history is experienced as the taken for granted, axiomatic necessity of objective reality. It is the foundation of the *habitus*". Further, Bourdieu (1968, p234) himself states that

> Culture is … achieved only by negating itself as such, that is, as artificial and artificially acquired, so as to become second nature, a *habitus*, a possession turned into being … culture is not what one is but what one has.

These statements are instructive and illuminating in terms of the case of Liverpool. *Habitus* needs a long time to take root, so its origins appear to be historical when they are identified; an appearance which lends gravitas and weight to the phenomena which the habitus validates and supports. Culture, meanwhile, plays a significant role in the collective or group habitus, in that it is one of the "structuring structures" (Johnson, 1993, p5) which binds people together and promotes consensus; further, the notion that culture is something that one *has* rather than something that one *is* reflects Bourdieu's concept of *social capital* – whereby status and power are more valuable than material wealth (1985). Liverpool's contemporary focus on history, culture, and status mirrors the history and culture that the *habitus* needs for it to become established, providing further justification and validation for the position that heritagisation and commemoration are the best, if not only, option for the continued regeneration of the city. Thus, *habitus* as a process merges with its goal of creating popular and widespread acknowledgement that Liverpool's history, as embodied in the built environment of the city centre and the waterfront, rendering it further imperceptible and, therefore, more effective.

The concept of home is nigh on a universal one, in that the vast majority of humankind have an understanding of what a home is, be it positive or negative, physical or spiritual, experiential or abstract (Hollander, 1991; Tamm, 1991; Somerville, 1992; Weiman and Wenju, 2007; Tete, 2012). Because the concept is ubiquitous and fundamental to many aspects of social interaction, social divisions, structural location, and self-identity, it is vital that the academy continues its exploration into it. Simultaneously nebulous and self-evident, omnipresent yet individualised, home as a concept remains a foundational element of all social study. Projects such as this further the existing body of understanding of the social world and may play some role in challenging the unequal power relations underpinning the vast range of experiences of home, or lack thereof.

The shift of emphasis from the immediate home and neighbourhood, to the city centre and waterfront, that is identifiable from the data, is one that appears to go unnoticed in other contemporary literature and commentary on regeneration initiatives in Liverpool. Further, none of my respondents appeared to acknowledge or recognise that the ideas they voiced about regeneration and home in Liverpool were markedly different according to era. This may suggest that there is a degree of consensus that contemporary prioritisation of the city centre and the waterfront is both appropriate and desirable for all. However, *habitus* operates in the shadows; it functions on an unseen, and therefore unnoticed, level – those influenced by it are unaware that their world view has been manipulated, and unaware that their world view is one shared by many others in the same social class (Bourdieu, 1977). Thus, *habitus* in this context encourages support for prioritisation of the regeneration of the centre of Liverpool and discourages any objection which may lead to a curtailment of the city centre project in favour of a re-prioritisation of the districts. The *habitus* ensures that policy makers, politicians, and architects of governance do not need to seek overt approval for their strategy of nurturing and restoring the city centre at the expense of the districts. Rather, tacit acceptance is achieved via widespread, stakeholder-style acceptance that the centre and waterfront are the Liverpool home that must come first.

As presented in Chapter One, I have been able to discern different stances or approaches when describing, exploring, and theorising the various aspects of Liverpool life and phenomena. Using the titles I employed to signify these distinct approaches, I hope that what I provide here contributes a new facet of understanding to what I earlier termed the 'mardy' standpoint, which acknowledges harms and detriment to the city and seeks both recognition of said harms, and redress. In terms of the existing body of literature and associated academic thought on the concept of home, my book constitutes a call for further exploration and theorisation of home as community, neighbourhood, or district: The equivalent of *baille*, *mahalla*, and *marae* without the cultural specificity, and identification of a term that could sit between *heim* and *heimat*, bridging the gap between and complementing the meanings of each.

This project has at its core qualitative primary research conducted with 30 people who acknowledge Liverpool as their home in one form or another. As such, it is inherently subjective, and could never be 'verified', or provide a means of generalisability in any scientific sense. However, the project is much more about concepts, ideas, meanings, and feelings than it is about patterns or trends; it does not purport to be neutral and seeks neither to prove nor generate any hypothesis. As established in Chapter One, home as a concept is deeply entrenched in human emotions. In the case of Liverpool, home emotions have been manipulated and exploited in the furtherance of the capitalist goal, and can be seen as part of a much wider, global capitalist process of emotional manipulation and coercion as identified by Mestrovic (2015). Findings such as those presented here provide further evidence of these processes, and possible ways to make sense of them.

Just as both Liverpool people and the wider world have been continually pushed to understand Liverpool as a city to be different to the point of uniqueness – for both positive and negative reasons – Liverpool's citizens are also constructed as different from other populations. The stereotypes that have been discussed throughout this project collude with other social processes and reputations to portray the Scouser as uniquely bad: Criminal, feckless, immoral, and litigious; other processes and conceptions meanwhile present the Scouser as uniquely good: Glamourous, funny, and ready to party more so than any other social group. In reality, however, this data indicates that, rather than being inherently distinctive, Liverpool people are ordinary – qualitatively no different from the population of any other working-class, post-industrial city. My respondents are all 'good' people, with interesting personalities and interesting life stories, and all are motivated by family and friendship bonds, both in spite of, and as a result of, the variety of structural factors and processes that the city has been subjected to.

Finally, and on a note which combines the personal with the academic, the project has illuminated processes of change in the Liverpool area, some of which are destructive, and some which are evidential of a domicide. One area where this is apparent is the Scotland Road area which, while mismanaged, misrepresented, and misunderstood for decades, was arguably dealt its terminal blow with the construction of the second Mersey tunnel, which rendered the neighbourhood unrecognisable. As the 50th anniversary of the opening of the Kingsway tunnel approaches in 2021, the time might be right to build on the ideas developed in this book and design a further qualitative research project reflecting on the impact of the tunnel's construction on the Scotland Road community, and families like my own.

Bibliography

Bourdieu P (1968) 'Outline of a sociological theory of art perception' in Johnson R (ed) (1993) *The Field of Cultural Production: Essays on Art and Literature*. Polity: Cambridge, pp 215–238.

Bourdieu P (1977) *Outline of a Theory of Practice*. Cambridge University Press: Cambridge.

Bourdieu P (1983) 'The field of cultural production, or: The economic world reversed' in Johnson R (ed) (1993) *The Field of Cultural Production: Essays on Art and Literature*. Polity: Cambridge, pp 29–73.

Bourdieu P (1985) 'The genesis of the concepts of habitus and field' *Sociocriticism* 2 pp 11–24.

Crick M (1997) *Michael Heseltine: A Biography*. Penguin: London.

Hollander J (1991) 'It all depends' *Social Research* 58 (1) pp 31–49.

Jenkins R (2002) *Pierre Bourdieu*, 2nd ed. Routledge: London.

Johnson R (1993) 'Editor's introduction: Pierre Bourdieu on art, literature and culture' in Johnson R (ed) *The Field of Cultural Production: Essays on Art and Literature*. Polity: Cambridge, pp 1–28.

Mestrovic S (2015) *The Post-Emotional Bully*. Sage: London.

Somerville P (1992) 'Homelessness and the meaning of home: Rooflessness or rootlessness?' *International Journal of Urban and Regional Research* 16 pp 529–539.

Tamm M (1991) 'What does a home mean and when does it cease to be a home? Home as a setting for rehabilitation and care' *Disability and Rehabilitation* 21 (2) pp 49–55.

Tete S (2012) '"Any place could be home": Embedding refugees' voices into displacement resolution and state refugee policy' *Geoforum* 43 pp 106–115. doi:10.1016/j.geoforum.2011.07.009.

The Lego Movie [DVD] (2014) Chris Miller Phil Lord dirs. USA: Village Roadshow Pictures, Animal Logic, Warner Bros Animation.

Weiman M and Wenju S (2007) 'Home: A feeling rooted in the heart' *Children's Literature in Education* 38 pp 173–185. doi:10.1007/s10583-006-9023-3.

Pen portraits

Mary

Mary is my closest friend. We met at secondary school and were friends from the start; although our friendship deepened during our time in sixth form and cemented when we were in our twenties. Mary is seven weeks younger than me. I was maid of honour on her wedding day and I am her youngest son's godmother; she is my son's godmother and was maid of honour at both my weddings. At the time of the research, Mary lived alone as a result of her marriage breakdown, and her two adult sons were away at university. She is a primary school teacher, specialising in early years, in a school in Kensington.

Rita

Rita is Mary's mother-in-law, with whom she remains very close in spite of her separation from her husband. I have known Rita since 1992. Rita has, apart from a few months in Toxteth, spent the whole of her life living in Kensington, bringing up her three children and working locally, including at the famous Meccano factory where she took part in the 'sit in' in protest at closure. Rita lost her husband Eric in 2000, and her own health has declined steadily since. She leaves the house very rarely now but enjoys a steady stream of visitors daily.

Jill

Jill and I met at the University of Liverpool not long after I started this project. She completed her doctorate in 2016 and, at the time of the interview, was looking for work in her field of sociology and culture. Jill was 46 at the time of the research and lived alone in Wavertree, close to her brother Jack, who has Down's syndrome and lives in supported accommodation. Jill suffers from joint hypermobility syndrome which impedes her mobility.

Irene

Irene is Rita and Mary's Avon lady. She is a retired classroom assistant and used to work in the same school as Mary. Irene is very active in the Kensington community and has always lived in the area. Mary approached Irene on my behalf, as she felt that Irene would be interested in talking to me, particularly about the Kensington New Deal (KND). Irene lives alone, but has a very active social life, and has been single since her daughter was very young. Her daughter is now married with a daughter of her own and lives in Germany.

Jodie

Jodie is my father's great niece, i.e., she is my first cousin once removed. At 39 she was one of my youngest respondents. I approached Jodie to take part as (a) she has experience of, and an interest in, urban regeneration and (b) she is intelligent and interesting. My interviews with Jodie generated more data than any other participant. Jodie is a single parent to her son and lives on the same estate where she grew up.

Sandra

Sandra is married to my cousin Joey, who is my mother's nephew. Sandra and Joey had been married for 31 years at the time of the research, and they have two adult children. She was 52 years old and worked as a hospital office manager in Brisbane, Australia, where they emigrated to in 2005. Sandra grew up in Wavertree and then Huyton and lived in both Walton and Prescot before emigrating. As she lives in Brisbane, Sandra completed a questionnaire rather than an interview.

Joey

Joey is my cousin on my mother's side, and Sandra's husband. At the time of the research, he was 53 years old and had done many jobs, including working in the Merchant Navy, as a driving instructor and as a caterer. For several years now he has worked as a camera operator at big events and specialises in stadium tours. As is the case with most of my maternal family, Joey grew up in the Page Moss/Huyton area. Joey is nine years older than me; as is customary in big families, he was often charged with looking after me when I was small, and I have very fond recollections of him taking me out around the estate where he lived. Like Sandra, Joey completed a questionnaire rather than an interview.

Dave

Dave is also my mother's nephew, and first cousin to myself and Joey. He grew up in the Huyton area, before leaving Liverpool to work in Plymouth in his late teens. Before emigrating to Australia, and eventually settling in Galway, Republic of Ireland, Dave lived for several years on Hope Street in central Liverpool. Like Joey, Dave has worked mainly in film and television production; however, he now works as a life coach. He lives with his wife and his 12-year-old son. Like Sandra and Joey, Dave completed a questionnaire rather than an interview.

Susie

I have known Susie since we both started doing a Criminology Master's degree in 2002, although I have actually known her husband since primary school. I really wanted to interview Susie for my project because of the various volunteer and paid roles she has had over the years, which have included outreach with asylum seekers and refugees, homeless people, sex workers and survivors of both domestic violence and sexual assault/rape, all within a Liverpool setting. Data generated from my conversation with Susie was far reaching, as she was able to talk about her professional experience, her childhood in Beck Grove (Thornton) and her experience as an Aintree resident of managing/negotiating the Grand National event every April.

Martha

Martha was in the audience of a paper I gave at the University of Liverpool; she approached me at the end of the session to ask me more about my project. Martha was interested because the topic area chimes with her lived experience, as she, like me, took a long route into academia, and has a background similar to mine. Ultimately, she volunteered to be interviewed for this project and provided me with a compelling, inspirational account of her childhood in Cantril Farm/Stockbridge Village, and the path she has taken to her career in academia. Martha now lives in Rainford with her partner and two sons and was able to discuss responses to her and her family when she first moved into this area as a 'Scouser'.

Jackie

I first met Jackie at secondary school, where she was in my class and part of a loose circle of friends with myself and Mary. At the time of the interview she was 43 and lived alone with her dog in Fazakerley. Jackie has one adult son who is at university and she is very close to her mother and younger sister. She works in central Liverpool and has a deep concern for the people she sees who appear to be homeless or experiencing social/financial issues. The interview I conducted with Jackie and her mother Maureen was perhaps the most emotional of all, as a result of both the financial hardship they have endured and the personal tragedy that they have experienced.

Maureen

Maureen is mum to Jackie, a son and another daughter. At the time of the interview she was living in Fazakerley with her second husband, who, although still working, had been diagnosed with terminal cancer. Her relationship with her second husband was long, happy and secure, and characterised by financial stability, tenderness, love and fun – markedly different to her marriage to Jackie's father. Like Jackie, Maureen had extensive experience of loss in her life, but had reached a place of peace and contentment. Subsequent to the interview in July 2016, Maureen lost her husband to cancer and is facing the future with support from her children and grandchildren.

Ronnie

Ronnie was 62 at the time of the interview and had had a long working life in Liverpool, mostly in social housing and related areas. An alumnus of the University of Liverpool, he has extensive experience of housing patterns in the Liverpool area, and has been involved in several 'ground up' regeneration projects, including Granby Four Streets and Homebaked. As a direct result of interviewing Ronnie, the house that I grew up in was the first property to be renovated and rented out by Coming Home Liverpool, a social enterprise headed up by Ronnie, which seeks to bring empty houses in Liverpool back into use at a fair rent.

Lynne

Lynne is Rita's older sister. Rita put me in contact with Lynne and her husband Brian as they have lived for many years in Cantril Farm/Stockbridge Village. Now in her eighties, Lynne enjoys her retirement after having various part-time jobs over the years, and looks after her granddaughter, who stays with her while doing her degree at the University of Liverpool. Although she does not see them much anymore, Lynne remains very close to her siblings and speaks to Rita most days on the phone. Lynne is proud to hail from Kensington, but says she prefers shopping in St Helens rather than Liverpool these days.

Brian

Brian is Lynne's husband. The couple have been married for almost 60 years and have three children and five grandchildren. Brian had a difficult childhood in many ways and feels that his life really began after he met and married Lynne and became part of her family. During the early years of their marriage Brian went to sea for long periods with the Merchant Navy, later becoming a bus driver and delivery driver. Now retired, he enjoys his time with Lynne and is relied upon by the entire family for lifts. The joint interview I conducted with Lynne and Brian in December 2016 was perhaps my favourite; their love for each other, and delight in each other's company, was infectious, and their easy manner makes them excellent company.

Christine

Christine is my mother's niece and my first cousin. Christine's mother, my aunt, is 22 years older than my mother and Christine is only two years younger than my mother; thus, she feels and behaves more like my aunt than my cousin. Christine lives in Huyton with her husband and retired just before her 60th birthday, after working for many years for Barclays bank. Her two daughters – one a primary school teacher and one a legal executive, live close by, and she enjoys looking after her two small grandchildren.

Roy

Roy is Lynne and Brian's eldest son. Brought up in Kensington and subsequently Cantril Farm/Stockbridge Village, Roy now lives in Knowsley with his wife, a secondary school teacher, and his two teenage sons. After leaving school he joined the Royal Air Force and, on leaving, subsequently retrained as a marine biologist. Roy is passionate about education and holds several voluntary positions in school governance in the Knowsley area. In his spare time, he enjoys tinkering with old Range Rovers and tending to his pet hens. The interview I conducted with Roy in November 2016 was very enjoyable, not least because he has an amusing turn of phrase and is given to florid descriptions and expressive language.

Enid

Enid is the paternal aunt of a close friend of mine. She grew up initially in Fishguard Street, Everton, and moved with her family to Kirkby in the 1950s. As a young woman she emigrated to the USA, where she met and married her American husband. They now live in Virginia Beach where Enid works as a nail technician. Enid still has very close ties to her family in the Liverpool area, and she and her husband visit at least once a year, always staying at the Suites Hotel in Knowsley, where they first stayed when she attended my friend's wedding in 2003. As she lives in the USA, I asked Enid to complete a questionnaire; however, she preferred to write a statement, which I have taken as testimony.

Peter

Peter is my father's nephew and my first cousin. I approached Peter for an interview largely because I knew he had lived in a newly built 'Hatton' house from the 1980s onwards and because I know he likes to talk. Peter spent his childhood living in Vauxhall Gardens in the Vauxhall area of Liverpool, and had a difficult time at school, in part because of his severe childhood asthma. Like many other young men in 1980s Liverpool, Peter struggled for many years to get steady employment, and suffered from clinical depression as a result. As a talented actor, he picked up some bit parts in television programmes, but was unable to carve out a career in a notoriously precarious profession. He now lives with his partner and young stepson in the Anfield area and is a support worker with adults with learning difficulties. My interview with Peter was incredibly stimulating, as he was able to articulate many of the difficulties he faced as located against various social processes. I am proud to be related to an intelligent, astute, politically aware man like Peter.

Teresa

Teresa is the grandmother of a school friend of my son, and I was introduced to her as she had direct experience of 'slum' clearance as a child. Teresa spent the majority of her childhood in Pluto Street, Kirkdale, before moving with her family to Fazakerley as their house was earmarked for demolition. Now in her sixties, Teresa lives with her husband in Aintree, close to her two sons and their families, and her sister. Since her sons were little, she has worked part time as a health care assistant at Aintree Hospital (Fazakerley), where she works with people recovering with brain injuries. Among other things, Teresa spoke at length about the death of her mother during the interview in January 2017, and the difficulties she had felt dealing with her grief.

Tony

Tony is the father of a school friend of my son, and Teresa's oldest son. He was 39 at the time of the interview in January 2017, which took place in his home in Aintree. Tony has lived in the Fazakerley/Aintree area all his life, and now lives with his wife Jeni, his two daughters and his baby son. Like Peter, Tony had a difficult time at secondary school, but subsequently went to college and now works in railway maintenance. I asked to speak to Tony straight after my interview with Carl (see below), as I know him to be talkative . Of all my respondents, Tony asked by far the most questions and was very keen to understand the phenomena and processes that I was interested in.

Jeni

Jeni is the mother of my son's school friend, as well as Tony's wife and Teresa's daughter-in-law. I have got to know both Jeni and Tony quite well through my son's friendship with their daughter and, I discovered more recently, she is the first cousin once removed of Enid. Jeni was 39 at the time of the interview and pregnant with her fourth child. Jeni lived in Litherland as a child before moving to Maghull and spent a short period as a young woman living in Walton close to her paternal grandmother. She trained as a nurse when her oldest son was very small and, at the time of the interview, was a charge nurse, specialising in compliance/quality assurance. Jeni, Tony and Teresa are all wonderful company and I enjoyed speaking to all of them immensely.

Paul

Paul is the husband of a childhood friend of mine. Although he grew up in the Old Roan area on the Liverpool/Sefton border, Paul has lived in Huddersfield for almost 20 years. An accountant by trade, Paul now works in a senior role at a university in the north of England, and was working on a PhD thesis, which he has subsequently completed. Paul and his wife visit Liverpool regularly to see family and friends, and hope to move back to the area once he retires, as they have never felt 'settled' in Huddersfield.

Carl

Carl is a friend and work associate of my cousin Jodie. I met Carl at Jodie's house when I went to interview her the first time; he had popped in for a bite to eat. From a large Irish-Liverpudlian family, Carl spent his childhood in Bootle, and has lived in Jersey, Ireland, and Australia before settling in Southport with his wife and two teenage daughters. Carl can turn his hand to most things, but, in his sixties, he currently works as an IT consultant.

Anne

Anne is the mother-in-law of a close friend of mine. Together with her parents and many siblings, Anne lived over a bakery in Bootle as a child, before marrying her husband John in her late teens. She has had a variety of part-time jobs over the years, including when her six children were all small, and lived in Bootle, Seaforth and Aintree before settling in Maghull, where she still lives with John and her adult son who has special needs. Now retired, she enjoys a constant stream of visits from her children, grandchildren and great-granddaughter. Despite living away from Bootle for almost 50 years, she still misses the area and frequently visits friends there.

John

John is the father-in-law of a close friend of mine, and Anne's husband. As a child, John struggled at school, largely due to his poor hearing; however, he was later discovered while working on the docks to be a gifted mathematician, leading to him being promoted to office work which he ultimately rejected as he "couldn't stand being inside". John has done many jobs over the years and is proud of being able to turn his hand to most things, for example, renovating all of the houses he has lived in with Anne. Sadly, John passed away after a short illness in 2018.

Gary

I was introduced to Gary by my acupuncturist, Sarah, who is friends with Gary's youngest daughter. Aged 62, Gary had lived for the majority of his life in Speke, and for a short period in nearby Halewood and Gateacre. He currently lives in Speke with his second wife and their three adult daughters. The youngest of seven sons, Gary experienced financial hardship as a child, and had several jobs upon leaving school before joining the Royal Navy in his twenties. Gary has experienced poor health for several years now and had one of his legs amputated as a result. Despite his health, when I interviewed him in February 2017, he was considering embarking on a degree programme. Gary was welcoming and friendly when I interviewed him, and we had a very enjoyable conversation, punctuated by his dog's repeated and plaintive requests for food.

Pat

Pat is the aunt of a close friend of mine. She grew up in a tiny house in Wright Street, off Scotland Road, and was in her late teens by the time her family were rehoused in the Southdene area of Kirkby. Pat worked in the offices for a chain of local clothes shops and, when she married her teenage sweetheart Joe, set up home in Fazakerley before moving to her current home in Maghull. Pat has three sons and one daughter, and several grandchildren who visit her frequently. In 2010, Pat suffered the misfortune of losing both her husband and her sister within a week of each other; however, she remains positive, keeping in close contact with her family and enjoying regular outings to sing-songs in a pub in central Liverpool. I interviewed Pat in 2017, shortly before her 80th birthday.

Maureen

Maureen is Pat's younger sister, and my friend's aunt. She was in her early teens when the family moved to Kirkby and is proud to have passed the 11-plus and gone to Notre Dame Everton Valley grammar school. Like Pat, Maureen settled with her husband Phil in Fazakerley, and brought up her two sons and two daughters there before moving on to Maghull. However, only a few months after the house move, Maureen's youngest child Gary was killed at the Hillsborough disaster, and this tragedy has provided the context for much of the family's life since. Like Pat, Maureen is in close contact with her family and enjoys spending time with her grandchildren.

Colette

Colette is my friend's mother and the younger sister of Pat and Maureen. Colette was a baby when the family relocated to Kirkby and was the last of the eight siblings to be born in Scotland Road. Colette loved their house in Southdene and felt it was always a happy, cheerful house, but it was never the same after her brother Paul was killed while away at sea in the early 1970s. Colette's parents struggled to recover from the tragedy, although things did improve somewhat when they used compensation for their son's death to buy a bungalow in Maghull. Colette and her husband Bill settled in Maghull where they have brought up their three daughters – my friend being the oldest – and now enjoy the company of their four granddaughters and one grandson. Now in her late sixties, Colette is enjoying her retirement after many years working in clerical and receptionist roles.

Index